Landmarks in the History of Science

Great Scientific Discoveries from a Global-Historical Perspective

Basil Evangelidis
Leiden University, Netherlands

Vernon Series on History of Science

VERNON PRESS

www.vernonpress.com

In the Americas:
Vernon Press
1000 N West Street,
Suite 1200, Wilmington,
Delaware 19801
United States

In the rest of the world:
Vernon Press
C/Sancti Espiritu 17,
Malaga, 29006
Spain

Vernon Series on History of Science

Library of Congress Control Number: 2016959066

ISBN: 978-1-62273-249-4

Front cover illustration: The "Archimedes" Portrait was painted about 1620 by Domenico Fetti in Mantua. It probably remained in Italy until 1743. Today the picture is located in the art museum Alte Meister in Dresden (Germany).

Contents

List of illustrations

"My purpose is to set forth a new science dealing with a very ancient subject"

(Galileo, Dialogue Concerning Two Sciences, Third Day)

Foreword

The scope of this book is a short journey through the last 2400 years of consciously recorded scientific practice. From the aspect of this considerably long period of time (from Ancient Greek, to Chinese and Islamic Science until the Age of the Discoveries and Modern Science and Technology), the greatest advancements in the world-history of science may be found not only in the theoretical field, such as with heliocentrism, atomism, relativity, but, more important, in the methodological transition to the experimental, mathematical, constructivist, instrumental practice of science.

The advancement of science, from antiquity to the renaissance, was significant in the domain of medicine, especially in the anatomy, the pathology and the hygiene, which may be ascribable mainly to the physicians and anatomists Thaddeus of Florence, Mondino de Liuzzi, Jacopo Berengario da Carpi, Andreas Vesalius, Realdo Colombo, the tradition of the works of Hippocrates and Galen, and that of Muslim scientists such as Muhammad ibn Zakariya al-Razi. The ancient medicine, however, believed that the venous blood is generated in the liver, from where it was distributed and consumed by all organs of the body. Willian Harvey was the one who recognized the importance of the circulation of the blood, in his work *On the Motion of the Heart* and Blood. Harvey was also one of the first embryologists.

From the inner organs of the organisms to the outer regions of earth, renaissance science was ubiquitous. Significant discoveries were taking place in geography and cartography: The Norse voyages to Greenland and North America and the African travels of Masudi, Ibn Haukal, El-Bekri and Ibn Battuta had an inappreciable influence to Western Europe. A relatively larger impact had the journeys of John of Plano Carpini, William of Rubruck, Nicolo, Maffeo and Marco Polo, in the thirteenth century, and the voyages of John of Monte Corvino, Odoric of Pordenone, Andrew of Perugia, Jordan of Severac, and John of Marignolli, in the early fourteenth century. Aside from the eyewitness or hearsay story of Masudi, who "believed the 'green sea of darkness' (the Atlantic) to be unnavigable, and the frigid and torrid zones of the earth to be uninhabitable" (Parry, 1963: 5), Jewish cartographers and instrument-makers working in Majorca in the later fourteenth century, especially Abraham Cresques, produced, by about 1375, the famous accurate Catalan

Atlas. He applied, for the first time, medieval hydrographical techniques to the world outside Europe, representing places such as Timbuktu and the rivers Senegal, Niger, and Nile.

The Iberian Peninsula was a meeting point and crossroad of mutual affection between Arab, Jewish and European culture. Alfonso X of Castile summoned into his court intellectuals of three religions, his works were translated into French and his astronomical tables were annotated by Copernicus. Spanish culture was also influenced by the Arabs, in the vocabulary, in architecture, in commerce, irrigation, the design and rig of ships, in the construction of saddlery and harness. The Arabs were found to possess the original manuscripts of Greek scientists, which they translated and commented. When the Christians conquered the library of Toledo, they found numerous writings, while some searching for Ptolemy's *Mathematical Syntaxis* or *Almagest*. The intellectuals knew about that work, where Claudius Ptolemy exposed the geocentric system (based on observations made with naked eye).

The Toledo School in Spain, directed firstly by Archbishop Raymond of Toledo, hosted a significant movement of translators, during the twelfth and the thirteenth centuries. The practice of translation extended also to other libraries in Spain and locally organized translation workshops. The translators were doctors who served in the courts of the rulers and knew Judeo-Arabic and Latin. They were Jewish, Italian, such as Gerard of Cremona, baptised Jews, as John of Seville and Dominicus Gundissalinus, of Latin or other origins, such as Michael Scot and Rudolf of Bruges. There are two families of works translated from Arabic translations: a) Works concerning practical knowledge related to everyday life: The medical works of Hippocrates and Galen, various projects of mathematics that have particular utility as geometric works, reflections on ratios, the fifth book of Euclid, various books of Euclid's *Elements*, astronomical and engineering works, such as for pumping water, manufacturing catapults etc. b) The other family was favoured by medieval scholars, works mainly of Plato and Aristotle. Later, the contact of the Europeans with Greek texts was stopped and some translations are incomplete.

By 1200 in Paris, Bologna, and Oxford the students were hundreds, and learned liberal arts, medicine, theology, law, while from 1377 to 1520 more than 200,000 students passed from German universities. The universities had three courses: a) The faculty of arts involved internally two cycles (corresponding to Bachelor and Master). The student studied the trivium (grammar, rhetoric, dialectic) and the quadrivium (arithmetic, astronomy, mathematics, physics). b) The second cycle was standardized, biennial and the teachers were many. The students learned mathematics, natural philosophy, astronomy, music, metaphysics, poetry, and ethics. c) The doctoral cycle offered theology, medicine, and law and it was extremely long. In Paris, there were four schools, a graduate school of liberal arts and three postgraduate (law, medicine, theology) (Rüegg, 1993).

Navigating and trading resources

The Spanish invasion in North Africa began with the capture of Melilla in 1492. The next year Columbus reported of islands in western Atlantic and insisted that they might be used as stepping-stones to China. Meanwhile the art of printing made possible a diffusion of navigational manuals and spread the news of discoveries, with bestsellers such as Peter Martyr's *De Orbe Novo*, Fracanzano da Montalboddo's *Paesi novamente retrovati*, Sebastian Münster's *Cosmographia universalis*, Theodor De Bry's *Grands Voyages*.

Seaborne trade was traditionally organized by merchant guilds, craft guilds, regulated companies, as a type of commenda, societas, and compagnia. A fifteenth-century merchant ship might take up to two months to make the passage from Barcelona to Alexandria; perhaps two or three weeks from Messina to Tripoli; ten or twelve days from Genoa to Tunis. In the fifteenth century, the small Atlantic ships of Basque, Galician or Portuguese origin, invaded in the Mediterranean. The economic activity of the region was concentrated in Milan, the center of metallurgical industry, Florence, the main textile and banking center, Genoa and Venice, the centers of Eastern luxury trade to western and northern Europe. The Genoese capital associations *compere* and *maone* had a corporative character. Constantinople and Cairo were immense urban and consuming centers. Florence, Genoa, Venice, Ragusa, Naples, the western Mediterranean as a whole was rarely self-sufficient and depended upon sea-borne trade in grain, salt, food preserved in salt, oil, wine, cheese, raisins, currants, almonds, and oranges.

> *Towards the end of the century, however, exports of oil from Andalusia began to be directed to the Canaries, and later to the West Indies, where it commanded very high prices. The Mediterranean wine trade -since viti-*

*culture was spread throughout the region- could not
compare with the great fleets which left the Gironde,
and later the Guadalquivir, for Atlantic destinations*
(Parry, 1963: p. 39).

The far eastern trade was controlled by Chinese, who delivered
spices in the important Malayan port of Malacca. From there, together with the cinnamon of Ceylon and the pepper of India, the
spices were sold in the *spice ports* of the Malabar Coast and Gujarat.
From Malabar, Arabian teak-built ocean-going baghlas followed
two alternative routes from the Indian Ocean to the Mediterranean,
and two principal ports of transshipment: Aden to the Red Sea and
Ormuz through the Persian Gulf.

Chapter 1

An introduction to Ancient Greek Science

Alongside the works of physicians, such as Hippocrates and Galen, the Greeks gave birth to a series of seminal theoretical, physical and mathematical works, especially in philosophy, logic, geometry, trigonometry, mechanics and astronomy. A Dutch anthropologist, Johan Huizinga (1949) observed, however, that the Greek philosophy and science were not products of the school, but they were fruits of leisure and had a playful form. He pointed out that the sophistic movement was closely associated with the primordial game and the riddle. It is something like a swordsman's trick, as Huizinga wrote.

The Greek word πρόβλημα (problem), in the original sense, meant something that someone puts in front of him in order to be protected - a shield, for example - or an instigation that one throws at the feet of another man - the "glove" that symbolised a challenge. Both meanings, if they are abstractly examined, apply equally to the art of the sophist. The sophistic questions and arguments were mainly problems in this very concept. One of those puzzles of leisure (σχόλη) or the first school, accompanied by rewards and fines, were the following: "What remains the same everywhere and always?" "The time".

Our view is contrary to the above mentioned. Both rhetoric and science developed in Ancient Greece as organised educational activities, through numerous philosophical and rhetorical schools and institutions. Once the favourable conditions disappeared, the cultural and scientific production declined. This is clearly exemplified by a simple comparison: Between the Greek mathematical works that are today extant one may find the *Elements* of Euclid, various works of Archimedes, the largest part of the *Conics* of Apollonius, the biggest part of the works of Heron, the largest part of the *Mathematical Syntaxis* of Ptolemy, the *Synagoge* of Pappus and the biggest part of the theory of numbers of Diophantus. On the contrary, lots of Greek mathematical works are considered as perished: All the works of Eudoxus of Cnidus, all the works of Democritus, all

the writings of the Pythagoreans Philolaus, Thymaridas and Archytas, the biggest part of the works of Aristotle, many works of Euclid,
all the works of Aristarchus of Samos, of Eratosthenes and Hipparchus, many writings of Archimedes, Apollonius, Heron, Diophantus and many others.

The mathematical description of lever, the invention of the com-
pound pulley and the water screw by Archimedes, prove that
mathematics are inextricably connected with the miraculous problems of everyday technological practice. Alain Bernard (2003), following the opinion of Wilbur R. Knorr (1993), that the mathematical
practice is a key tool for reflecting and understanding ancient tradition in geometry, proposed to study the ancient rhetoric. Furthermore, Giovanna Cifoletti (1995) observed that the rediscovery of
classical rhetoric during the sixteenth century, it was critical for the
reorganization of knowledge at that time, in which algebra emerged
as a distinct science.

During the sixteenth century, the terms *problem* and *analysis*
were redefined with the study of Pappus, Diophantus and Proclus
and the medieval algebra. The term *problem* was gradually identified with the rhetorical concept *quaestio* (as *quaesita,* things sought,
are distinguished from *data,* things given), while the concept of
analysis was partially connected with the rhetorical *ars inveniendi*
and with symbolic algebra. The question asked by recent research is
whether the rhetoric and the analysis were from the beginning
linked together.

These inquiries are motivated by the fact that while the analysis in
the broad sense offers specific ways of finding (inventing) solutions
to problems, *invention,* on the other hand, was one of the five parts
of ancient rhetoric. Bernard presents the opposing views of Unguru
(1975) and Knorr (1993), concerning the role of mathematical practice in historiographical interpretations. He focuses on the concept
of analysis in antiquity and proposes answers based on the study of
rhetoric. Bernard observes that the formal part of ancient geometry
was articulated on the basis of the synthetic method, whereas the
heuristic part followed the analytical method.

Pappus, according to Bernard (2003), distinguished between the
problem, where an act or construction is being displayed, and the
theorem, where, by assuming some presuppositions, one may examine the consequences of all the occurrences. The controversy
between philosophers on this issue began with the mathematicians
Speusippos and Menaechmus in the fourth century BC. Further-

more, Euclid's works were the topic of commentary upon the introduction of this distinction by Proclus.

1.1. Plato and Aristotle upon truth and Ethics

*Plato's and Aristotle's seminal contributions are considered as foun-
dations of Philosophy, Ethics, Political Science and Philosophy of
Science, in their inextricable interaction. They have also influenced
everlasting disputes over methodology and worldview, which have
directed significant scientific disputes, along centuries.*

Plato is the founder of philosophical critique, engaging rigid
questioning on the acquisition of knowledge in civil society, science, specific arts and faculties. Pure and practical reason is inter-
woven in Platonic dialogues. Ethics, justice, wisdom, courage, tem-
perance, piety and, in general, virtues, in their ambivalent interplay
with scientific reason, are equally important in the query for Good
and Truth.

Plato formulated his style influenced by the Socratic maieutic or
obstetric method of didactic. The Platonic philosophy distinguishes
contradictive ideas that may govern and guide scientific research
and ethics. The emphasis is given on critique, disagreement, altera-
tion, discontinuity; in other words, upon defining substances by
differentiating. On the other side, apart from the Socratic method of
refutation (elenchus), Plato introduced the method of Hypothesis
in order to acquire the knowledge of an answer to a specific question, when no one who already possesses that knowledge is to be
found (Benson, 2003).

Plato did not consider rhetoric as a means of education because
he claimed that practical rhetoric was a refutation of justice. On
account of this, Socrates in *Gorgias* distinguished between sophistry
and true art of speaking. He also distinguished between rhetorical
persuasion and *didactic persuasion*, which is the profession of the
mathematician and combines persuasion with teaching.

Science differs from belief because science is about understanding the essence, while belief is a hypothesis we make for the genesis
of something. However, Plato was also influenced by religion: The
dialogue *Menon* or *On Virtue* exposes the theory of anamnesis (Allen,
1959), i.e. *recollection*, of purely metaphysical origin. Recollection is
the influence of the performance of a past logical activity (*Laws*
732b); an inherent quality of the soul.

In *Phaedon,* the immortality and the divinity of the soul and the reality of the objects of its knowledge are considered as the cornerstones of philosophy. Religion in Plato was accompanied by myths: In *Euthydemus* (285 a d) the reader can find the myth of the flaying of Marsyas by Apollo. Aside from myths, the *Symposium* offers evidence of matriarchy, with the protagonist Diotima explaining the nature of love. She is a respected peer, proving the strong matriarchal elements of local life in Mantinea (created by settlements of small towns and villages, where, probably, women had maintained more powerful social status). Diotima was characterized as equivalent with the perfect sophists (208c).

The *Symposium* was written to explain the outstanding meaning of love. The educated peers exercise their minds in an appropriate manner to resist transitory passion and petty ambitions. The task is, therefore, that the poets, the artists, the educators, more urgently, who will shape with their knowledge the perfect citizen (*Laws* 643d) and the legislators, should direct young people into virtue. The pedagogical instruction of children, as reflected in the *Republic* and *Laws* (which may be studied in parallel), may amplify the spirituality of the people, establish ethical principles and nurture creativity. Children should neither become self-indulgent nor humiliated by punishments, but instructed towards idealization. By this way, takes place the metamorphosis of love from the individual body and personal soul to the love of beauty itself. This refinement and idealization of love is possible only through intensive mental training and inspired by the vision of Good.

Mental strength is delineated nicely with a familiar metaphor in *Phaedrus*: the mind is the charioteer who controls the route of the chariot, whereas the unruly horses of emotion and desire have to be harnessed. Desire is filled up with hubris and arrogance, while emotion can be managed only with orders and reason (253d). Another platonic dialogue is the *Sophist,* where the Stranger explains the method of conceptual division (219b-220a), and along with Theaetetus, they rank sophistry into the hunting arts (231a-c). The dialogue finally summarizes (231c-e) the characteristics of the sophist. The sophists argued, in a standard manner, by leaning on some *common loci*: inspections, lies, paradoxes, solecisms, redundancies and misplaced naming (because the things are infinite and the names are finite, and each name refers to many things, one can easily end up in absurdities).

The common loci have their own ways: *despite the word,* i.e. depend on the language (homonymy, doubt, composition, division, prosody, and word figure) and *away from the word,* i.e. they are independent of language, such as ignoratio elenchi, per accidens, absolutely or not absolutely etc. (Aristotle, *On Sophistic Refutations,* 1-5).

Aristotle, the student of Plato, distinguished three levels of knowledge: scientific knowledge, dialectic question and rhetoric likelihood. The demonstration, in rhetorical speech, is based on proofs, and the evidence is of two types, a deductive one, the *enthymeme,* and an inductive one, the *paradigm.* Both rhetoricians and dialecticians try *to discuss statements and to maintain them, to defend themselves and to accuse* others (1354a). The dialectic, however, according to Aristotle, is the reverse of the rhetoric. The rhetoric distinguishes likelihood from what only seems as likely to. It does not prove what is certain, but it only shows what is possible. The dialectic distinguishes reasoning from what appears as reasoning. In this way, the dialectic orator is being distinguished from the Sophist rhetorician, because the dialectician speaks in accordance with the degree of belief, while the sophist's speech is optional. The dialectic must rely on plausible arguments, i.e. probabilities, while the rhetoric is based on the *inverse to the plausible* arguments, as Hermes' argument, that he couldn't steal Apollo's cattle since he was born only yesterday.

At the beginning of the *Prior Analytics,* Aristotle observes that the demonstrative proposition is *the assertion of one of two contradictory statements,* i.e. it proves the truth of a sentence and the falsity of its contradictory; while the dialectic proposition is a *question upon contradiction* without highlighting the difference between the contradictory propositions. The dialectic reasoning may be of two kinds: a) deductive reasoning and b) induction i.e. foray from the details into the universals. We formulate dialectical reasoning through: Propositions (moral, physical and logical), divisions, locating the differences and similarities. However, the dialectics are not aimed at truth as philosophy does, but only at belief.

The *thesis* on Aristotle may be a *paradoxical perception of the acquainted* persons, that is a paradoxical subjective conception, such as 'everything is moving,' as Heraclitus said, or 'all is one,' as Melissus argued. The *thesis* is a problem, according to Aristotle (*Topica,* A, 11, 18-36). *Problem* is a dialectical theorem that contributes either to the election or avoidance, or to truth and knowledge (*Topica,* 11, 1). On regard of this theorem either the majority of the interlocutors express neutrality, or oppose to the opinion of the wise men, or the wise men disagree with the majority, or one with another. Almost all dialectical problems are called *Theses*. The thesis is a problem without explicit evidence, i.e. without end (1357 b5-10). It is infinite because there is no adequate and decisive evidence.

Aristotle does not interpret the universals in an idealistic manner. The reasoning, according to Aristotle, is demonstrative whenever it confirms the universal by showing that the individual i.e. each example is *overt*, obvious. Actually, Aristotle, in the introduction of the *Topica* distinguishes: a) *evidence*, derived from true and self-sufficient premises, from b) the dialectical syllogisms that rely upon *beliefs*, i.e. possible premises. Aristotle also notes that in pugnacious reasoning, the rhetorician takes only what appears as possible and what is contradictory to the possible; when, however, one expresses geometric aberrations, he doesn't rely on what appears to be possible, but on premises that do not conform to the *term*, i.e. to the definition.

Aristotle calls middle term of the premises of an inference the term which is entirely part or subset of a major term, while the minor term is entirely part or subset of the middle term (*Prior Analytics*, A). Aristotle points out that science does not proceed with this method, but directly, without middle terms. What he means is that science uses evidence grounded on the *tokens* demanded by the demonstrative inferences, namely the premises of scientific knowledge must be "true, primary, immediate, better known than and prior to the conclusion, which is further related to them as effect to the cause" (*Posterior Analytics*, B, 2, 2). This evidence is contrasted from the virtual proof of the contemplative (enthymemes) and dialectical propositions. The difference between science and belief is the following: Science arranges the universals and the necessary, which -we can show that- they are not contingent. Demonstrative reasoning, on the contrary, is based on evidence and therefore, because it is true, it is unresolvable, e.g. 'he has got fever, therefore he is ill' or 'she has given birth, therefore she has milk.' Every non-demonstrative reasoning is resolvable, e.g. 'he is breathing heavily, therefore he has got fever.'

1.2. Scientific topics in antiquity: measurement, experiment, and construction

The argument that we intend to support consists of the delineation of the organizing principles that created science and technology in antiquity. The answer that we give, as a challenge to research, is that there were three systematic basic sources of scientific advance: Measurement, Experiment, and Construction.

Construction was efficient in engineering manuals such as Heron's Pneumatica, with illustrations that depicted and represented the significance of the intuitional, synthetical and analytical procedures to proof. Measurement through construction, experiment and observation became exact after Archimedes' and Heron's approaches, which enabled a physical science that claimed to become complete, as it included, for the first time, the trial of fluids, liquids, air, and gases. Gas was particularly considered as a source of energy, with a whole series of steam-engineering applications, e.g. automatic doors, solar fountain, aeolipile, dancing automats etc. (Fabre, 2008). In fact, construction was so excessively important for Ctesibius' inventions and Heronian science that some historians wonder whether they should consider Heron as an artisan rather than as a mathematician (Vitrac, 2008).

Construction was a substantial practice for the proliferation of the topological categories. Heron's and Archimedes' works belong to the magnificent paradigms of the constructivist transformations of space. With their projects, the Mathematicians and Engineers Archimedes and Heron emphasized the significance of the synthetical and analytical demonstration, in practice, by constructing and joining vessels, pots, pipes, spheres, funnels, spouts and pulleys for testing important physical properties such as compression, pressure and force (*Heronis Alexandrini opera quae supersunt omnia*, 1899-1914; Lazos, 1995).

When science offered the means for the measurement of acreages, cylinders, and hemispheres, altogether with the representation of the proofs for their theories, then space became rationalized and subtilized. However, these scientific discoveries presupposed wide-ranging exchanges between civilizations.

The development of the numerical symbolism and the practice of cataloguing, in Egypt, Babylon and Greece (for example the catalogues in Parthenon, Athens), gave the opportunity for the formation of complicated accounting systems, which facilitated geometrical and astronomical discoveries. Landmarks in this evolution were the learning of the multiplication and division, the study of the assemblages and the partitions of fractions, and especially the sexagesimal numerical system of the Babylonians, whose utilization was critical for the measurement of cyclical, spherical, curved and celestial distances, surfaces and volumes (B.L. Van der Waerden, 2003).

In parallel, the diversity in the origins of scientific discoveries was represented not only by the Babylonian sexagesimal numerical system but also by the unprecedented contribution of India in arithmetic, with the introduction of the positional numerical system of the well-known figures: 1, 2, 3, 4, 5 ... etc. which simplified accounting as never before.

The significance of symbolism

Babylonians, Chinese and Greeks, with their preference to different numerical systems, tended to organize learning and spatial realities according to separate and distinct priorities. Thus, Chinese have shown preference to the picture, the Greeks categorized the quantities with the use of letters, either acrophonic or alphabetical, while the Babylonians insisted on the succinct semantics of the sexagesimal numerical system. It is no coincidence that the sexagesimal system has survived until today, through trigonometry and Ptolemy's astronomical books. The Pythagoreans also focused on symbolism, while they transferred significant truths and definitions from Ionia to Italy, such as that of the point as a unit having position, the distinction between right, obtuse and acute angles, the Pythagorean Theorem, the extreme and mean ratio, the regular polyhedral solids, the arithmetic, geometric, harmonic and musical ratios etc.

Philosophers contributed also to the scientific analysis of space and its properties. One of the first important discoveries, was Thales's proof that a triangle inscribed in a hemicycle is orthogonal. When Anaxagoras proposed that the formation of the world began with a vortex set up, in a portion of the mixed mass in which 'all things were together' (Heath, 1921: p. 172), he exposed a very early precognition of modern physical theories. At the same time he was offering a materialistic account of the universe, the sun, the moon and the heaven, insisting that the sun was a red-hot stone and the moon similar to earth, thus not gods, but matter tore away from the earth by centrifugal and centripetal forces.

Similar approaches stimulated the advancement of learning, ga-thered students and teachers in schools and opened the way for the discussion of special cosmological problems such as the cincture of the zodiac circle, in other words, the obliquity of the ecliptic. Later, the significant propositions stated by Democritus concerning the relations between the volumes of cones, cylinders, pyramids and prisms, altogether with the discovery of the quadrature of the lune by Hippocrates of Chios were signs that geometry and space meas-urement were meant to develop much more important theories.

Ctesibius, Heron, Philon of Byzantium, Archimedes, Plinius, Vitruvius, presented and constructed various inventions, such as pressure pumps, cochleae, hauled-bridges, mills, catapults etc. With appropriate equipment such as the lever, Archimedes discov-ered the laws of equilibrium, hydrostatics, and statics. The instru-mental aspect of the scientific exploration of space was intertwined with the rational and the mathematical. The ancient scientists used various instruments, such as sundials, celestial spheres, dioptres, the meridian circle, the astrolabon organon, the parallactic instru-ment and the mural quadrant (Sarton, 1959).

Euclid worked in Alexandria under Ptolemaios I and he was the head of the Mouseion. Among the most fruitful discoveries of Euclid (ca. 300 BC), we can admit, firstly, the introduction of the measurement of triangles and angles by their comparison to right angles; secondly, the propositions for the equality of the angles, on account of their parallel or alternative position; and thirdly, the proposition for the equality of triangles in terms of equal sides and angles (*Euclidis Elementa*, 1883-1884; Dodgson, 2009).

The special problems

The theoretical advancement was exemplified by the formulation of various methods (synthetic, analytic, reduction ad absurdum, in-duction) for proofs; while the investigations included three types of problems, according to Pappus: *Plane* problems, if they could be solved by means of the straight line and circle. *Solid* if they could be solved by means of conic sections. *Linear* if their solution required the use of more complicated curves, such as *spirals, quadratrices, cochloids, conchoids, cissoids*, etc. (Heath 1921: pp. 218-269.).

The squaring of the circle

Aristophanes and Antiphon of Athens were among the first Greeks who emphasized this problem. Later, Iamblichus, Archimedes, Nicomedes, Pappus, Proclus and Hippias of Elis contributed to the main scientific disputes, which were pivotally concentrated in the issue of the quadratrix and in Archimedes' Measurement of a Circle. In *Prior Analytics* (25, 30) Aristotle gave, as a counterexample of demonstrative inference, the squaring of the circle: our inability to square cyclical sectors by the use of menisci, results from the fact that it is not enough to bring one or a few segments, each being average of both cyclic and linear segment.

The trisection of any angle

The significance of this problem emerged from the need to construct regular polygons with nine, or any multiple of nine, angles. The reduction of that problem to a problem of conic sections was made possible by the use of inclinations or nutations (νεύσεις).

The duplication of the cube

Hippocrates of Chios, Eratosthenes, Archytas of Taras, Eudoxus of Cnidus, Menaechmus, Plato, Archimedes, Apollonius, Nicomedes, Heron, Diocles and Pappus had tried to solve this problem.

Pappus and all writers of late antiquity appreciated excessively the heuristic methods. Pappus also used demonstrative methods, public exhibition of a problem, and discerned the projections and the problems from the theoretical and stochastic investigations. The problem maintains rhetorical significance because it means dealing with a case as alleged.

The solution of a problem is invented in the process of finding, by reference to the loci and the appropriate locus resolutus, the analysed topic. The concept of analysis is linked to the concepts of game, play and study, as well. The study of the concept of learning course, *mathema*, was a classic topic not only in the platonic Menon, but also in sophistics, rhetorics and practice. The courses were not seeking so much for the purpose of persuasion as that of teaching. Gorgias insisted that rhetorical speech facilitates learning. In ancient times, in general, the questioning projections were usually the compendia, which were considered either as *problems*, if they contained some form of challenge, or as *assumptions*, if they were serving the purpose of the study of an initial challenge, e.g. the *Stoicheioseis* of Nicomachus and Boethius.

Reviel Netz (1998) observed that conservatism and classicism of the late antiquity do not mean degeneration. It seems plausible that in the time of Geminus (first century AD) the Greeks had formulated precocious theories of relativity, by considering that the sun, the moon, and the five planets move in a direction opposite to that of the universe. With the accumulation of scientific discoveries, economic exchanges and geographical explorations (corroborating Pythagoras's conception of a spherical Earth) three continents appeared:

Europe, Asia, and Africa, an encircling ocean, the Mediterranean, the Black Sea and Caspian, the Red Sea and Persian Gulf, the South Asiatic, and North and West European coasts were indicated with more or less precision in the science of the Antonines and even of Hannibal's age. Similarly, the Nile and Danube, Euphrates and Tigris, Indus and Ganges, Jaxartes and Oxus, Rhine and Ebro, Don and Volga, with the chief mountain ranges of Europe and Western Asia, find themselves pretty much in their right places in Strabo's description, and are still better placed in the great chart of Ptolemy. The countries and nations from China to Spain are arranged in the order of modern knowledge (Beazley, 1895).

1.3. Mathematical Astronomy

By 300 or 400 BC, astronomers were able to predict accurately lunar and planetary phenomena. With the exception of the contentious issue of "Phaenomena", the Euclidean was a plane and solid geometry, not a spherical one; that is to say, it was not applicable to the celestial sphere, as Autolycus' work "On the rotating Sphere" (at the end of the fourth century BC) was.

Greek Astronomy began with the organization of fixed stars into constellations, as Goldstein and Bowen (1983) supported. Apollonius, around 200 BC, proposed models of eccentric or epicyclical circular motions, while Hipparchus (about 150 BC) observed that the actual motions were much more complicated (Neugebauer, 1985). In the 4th proposition of Autolycus's project, we find the definition of the bounding circle, or else the *horizon*, i.e. the circle which separates the visible points of the sphere from the invisible ones (Rosenfeld, 1988).

A systematic account of spherical geometry was given by Theodosius (ca. 160 BC – ca. 100 BC). The majority of his propositions in
Sphaerics were stereometric. The role of geometrical construction
was emphasized by Theodosius. However, Theodosius' spherical
geometry was based on celestial realities, such as the special small
cycle of the circumpolar stars that never set, as they appear to an
observer on earth. The pole of this cycle is the celestial pole. A principal cycle that is tangent to this cycle and inclined to the parallel
cycles is the cycle of the horizon.

The parallel cycles represent clearly the phenomenal movement
of the fixed stars during everyday rotation, having as their pole the
celestial pole. A second major cycle, inclined to the parallel cycles,
is the zodiac cycle or ecliptic. The biggest parallel cycle is the equator (Spandagos, 2000). Theodosius' successor, Menelaus, at the end
of the first century AD, had been the indispensable *messenger* of
Spherical Trigonometry; while a completed system of Mathematical
astronomy was presented by Ptolemy (about 150 AD).

The Roman, the Persian, the Hindu, the Islamic and the Byzantine
Astronomy[1] represent the next stages in the dissemination of
science across countries and civilizations. The seeds of modern
theories were already present in Maimonides' (1135-1204) insistence upon the nature of light and design, while stressing the du-
bious parts of the Ptolemaic world-picture: "There is a difference
among ancient astronomers whether the spheres of Mercury and
Venus are above or below the sun, because no proof can be given
for the position of these two spheres" (Maimonides, n.d.: p. 289);
adding that "we adopt, in reference to the circuit of the sun, the
theory of excentricity, and reject the epicyclic revolution assumed
by Ptolemy" (Ibid).

---

[1] O. Neugebauer (1975: 11, 13) mentions the works of Gregory Chioniades,
   Gregory Chrysokokkes, Nicephoros Gregoras, Isaac Argyros, Theodoros
   Meliteniotes, Michael Chrysokokkes etc.

The transformation achieved

In the Roman Empire, with the works of Vitruvius (1914), the scientific conceptions for space became more concrete and analytical, regarding the profession of the architect, the building materials, the symmetry, the orders, the classifications of the buildings, the climates, the colours, the astronomical contributions to architecture, the machines and the implements, etc. In addition, Vitruvius offered precious information for the Greek philosophers and scientists.

It is evident, that space, knowledge, and power were interconnected in all the aforementioned scientific approaches. The experience of space is always socially constructed (Gupta and Ferguson, 1992), as shown by international exchanges and intercultural communication. Nodal transportation and commercial centres, such as *Trabzon* or the *Kingdom of the Cimmerian Bosporus*, had a transitional role for the relations between civilizations. The transcultural interdisciplinary perspectives were gradually enhanced, especially in Late Antiquity.

Therefore, there is a dynamical interconnection between the many different disciplines in their study of space, while enhancing their results by cataloguing, tables and axiomatization. These zestful discoveries were valuable enough to carry the quest out, until its completion.

1.4. Rhetorical and political sites

The use of the words topos, locus, place, site, was multiple: in the state language of the Hellenistic period the word topos was found in the Seleucids, referring to the territories of the cities. The Seleucid cities were excluded from the domination of the monarch, as only the countryside was a royal domain, where nations lived. The ruling class of the cities all over the empire, had to pay the well-known liturgies, which survived until the Byzantine period, and rhetoric was helpful in claiming or exemption therefrom.

The liturgies were permanent, such as choregy, hestiasis, architheoria, gymnasiarchy etc., and exceptional, like the introductory offering of unforeseen military spending and the trierarchy, which, because it was the most expensive, one could avoid it with the antidosis. Generally, the Seleucids respected the autonomy of the cities, as long as they did not undermine them and were kept in coalition with them. Later, at the times of Septimius Severus and Caracalla, in 190 AD, Julia Domna of the Seleucids gathered around her philosophers and sophists: Philostratus the biographer, the lawyers Papinian, Ulpianus and Paul, the historians Cassius Dio and Marius Maximus, the physician Galen and many other intellectuals and poets.

In the Hellenistic times the *grammatistai* provided the basic education, whereas the *grammaticoi* offered secondary education. The rhetoric was one of the faculties of higher education, along with medicine, philosophy, and engineering. The teachers of rhetoric and geometry, and the *grammaticoi* of the secondary education were privileged, compared to the primary *grammatistai*. Moreover, according to Lactantius (Inst. 3.25), while neither slaves nor women nor the poor could be deprived of music, geometry, astrology and philosophy, nevertheless, the dialectics, the grammar and rhetoric were distinguished by a different *societas*. It is no coincidence, that the Byzantines felt that the latest arts were essential to the *Eparchicoi*, the administrators.

Nevertheless, the Epicureans ("live hidden") and the Stoics had begun to suggest the withdrawal from the common affairs and undermined the city. Homeland was not the city, but the community for the Epicurean, the world for the Stoic. The Stoics in their works illustrated the ideal ruler. With this approach, they prepared the ascension of Marcus Aurelius, the emperor-stoic-philosopher. The political site is now considered as threatened and the orator is fastened anxiously to the city, but he is also interested to salvage much narrower interests or to benefit from the collapse of accumulating privileges, acquiring land (feudum, feud) and offices. The orators and the philosophers in this context acquire new roles.

This change is uncovered by the Athenian embassy of the philosophers Carneades, Critolaus and Diogenes to Rome in 155 BC and the embassy of the Rhodians and Attalus I in 201 BC. The cities Cyzicus, Miletus, Pergamum, Athens and the remote peninsula of Crimea (47 BC) were sending embassies for their autonomy and in order to avoid the fate of the cities destroyed by Demetrius the Conqueror, or the fates of Rhodes, Corinth, Xanthos and other cities that were destroyed during Roman expansion. Field of exercise of the rhetoric, as indicated by the dramatic orations and embassies during the intervention of the Romans, were the Κοινόν (Achaia, Aetolia, Stratonicea, Minor Asia, Pentapolis of Cyrene and the Panellinion that united cities of Africa and Minor Asia). In Kos, a grammarian, Nicias, was appointed as tyrant (41-33 BC) by Anthony.

Although the loss of the autonomy of Marseille in 49 BC created the impression that the course of events was not reversible, the emperor Nero, pressured by Gallic insurrections, declared Achaia, namely the whole of the metropolitan Greece, free (67 AD). The Romans founded Nicopolis and colonized Corinth (Laus Iulia Corinthiensis). Patras developed and dominated in the productive sector and, in the second century AD, Smyrna was considered as the metropolis of Greek intellectual life. Hadrian favoured the Greeks and built numerous buildings in Athens. The Athenaeum in Rome aspired to emulate the Mouseion of Alexandria, while during the second century AD the Mouseion of Smyrna and Ephesus stand out, as cultural centers, where eminent grammarians, rhetoricians, and physicians taught their lessons (Bengtson, 1991).

1.5. Alexandrian Renaissance

This chapter examines the foundation of scientific organizations in Alexandria, as a result of the communication and the linkages between civilizations. The benchmarks of that concentration of power and knowledge were mainly the advancements in geography, mechanics, architecture and geometry.

The confidence for the infinity of space altogether with maritime exchanges and colonisation must be seen as essential prerequisites for that advance. Excellent masters and builders, such as Deinocrates of Rhodes and Sostratos of Cnidus, and scientists, such as Strabo and Eratosthenes, confirm with their contributions the significance of this unperfected effort. Historical quest aims at the research of the interdisciplinary and multicultural contributions to the building of the organizations of the Alexandrian Library and the Mouseion. This study includes an approach to divergent theoretical and empirical conceptions of space implemented in antiquity, such as with the foundation of geometry and geography.

Place, order, and oscillation are significant instances of the categories involved in the spatial turn of the Kulturwissenschaften, as Gräßner (2011) insists. Clearly then, historical research should include a meta-theoretical recollection of various geographical, economic, social, cultural, intellectual processes and disciplinary approaches, for instance, numismatics (Berthold, 2011); whereas the interplay between knowledge and space is always dynamic.

In ancient civilizations, the notion of space was related to that of Heavens. The Greek poet Homer could scarcely think beyond the eastern Mediterranean and closely bordering lands, although he mentioned the river Thames. Homer also believed that the Ocean is a stream flowing around the disk of Earth. Not only Parmenides but also Aristotle insisted on the eternal, changeless, unitary, motionless conception of space (Ashtekar, 2006). Thus, the adventurous route from religion to science had to step through Astrology and the Zodiac, achieving by observation and measurement successive improvements in a long period of time.

Since "the mature Iron Age, maritime confidence, together with social and economic drives, propel Phoenician and Greek colonisation waves throughout the Mediterranean and the Black Sea," as Bintliff (2012: p. 2) repeats. Greater achievements in navigation came along with the voyages of the Carthaginian Hanno (about 500 BC), which opened the Strait of Gibraltar and the West African coast to the explorers. Further geographical evidence was acquired by Pytheas of Marseilles, who coasted the British islands and learned of northern islands beyond them. Almost simultaneously, Alexander the Great pushed the borderline between civilizations eastward to India and central Asia (Nowell, 1954).

The Mouseion (Μουσείον) of Alexandria

The Ptolemaic state was more coherent than the Seleucid, based on a Greek-Macedonian ruling class. Only a few cities were founded, such as Ptolemais, administrative centre in Upper Egypt (Thebais). In the old metropoles and the most distant settlements of Egyptian counties, Hellenic high schools were established and Hellenistic culture spread throughout the Nile valley. The state was highly centralized. There was the court of *brokers* for the Greeks and the court of *laokriton* for the Egyptians. Later the generals of each county acquired the right to try.

At the end of the reign of Ptolemaios I the Savior (323-285 BC), the Mouseion, a remarkable scientific research institution was founded in Alexandria. The director of the Mouseion, Demetrius Falireus, had a background as overseer of Athens and personal acquaintance of the function and composition of the Lyceum and the Academy. As he was a student of Aristotle, the greatest collector of books up to that time, it is likely to have been taught by the philosopher on how to organize a library. Perhaps for this reason, Strabo writes that Aristotle was considered as the spiritual father of the Mouseion and the Library. The dream of Demetrius started around 300 BC, with the construction of the *Mouseion*, the first university in the world. The Mouseion (*Μουσείον*) was a school established in accordance with the standards of the two Athenian schools, the Lyceum and the

Academy. It was given this name because it was dedicated to the nine Muses, the deities of the arts and sciences.

Strabo informs us that the Mouseion had a platonic type platform, where seats were preordained to gather philosophers. Next to it, there was the Aristotelian type of promenade for the debates between the scientists. Those who lived in the Mouseion were entitled to nourishment and payroll, their costs were assumed by the royal treasury, while they enjoyed tax exemption. The Mouseion was the upshot of a long tradition of systematic cultivation of the sciences by the Greek states. Two gigantic libraries, a troupe of Muses, comprised by scientists and artists, an observatory, an anatomical institute and a zoo, were included in the cultural system of the state. The Ptolemaic Egypt was the first state to provide organized healthcare for the population.

Heterotopia

The municipal leaders of Alexandria came from the class that maintained political privileges, those "from the Gymnasium", mainly burdened with the various liturgies. The gymnasiarch became the most important official in the Roman-occupied Alexandria and the Gymnasium was still the focus of the life of the Greeks. In towns, however, the Gymnasium was extinguished with efforts of the Romans. In 37 AD the Greeks were 6475 in Fayoum and 180,000 in Alexandria, but they had no opportunity for political action. The absolutism of the successors of Tiberius revolutionized both Roman senators and the Greeks, under the necessity to defend their libertas and prevent the crowding of Alexandria. Their struggles were also connected with the defence of the parliament, which was abolished by Octavian. Others argue that the parliament was stopped by the Ptolemies.

The building of the city of Alexandria and its respectful organizations, recalls questions about the construction of place as heterotopia, in Foucault's (1984) terms. Exchanges and colonization were parts of that creative movement. The wondering about the Unknown was a natural stance, always accompanying the discoveries and the centrifugal political movements. The emergence of the educational centres of Rhodes and Pergamon related to the exile imposed by Ptolemaios to the intellectuals, during the civil war of 144-145 BC. The Pergamon library had collected 200,000 rolls and the city was famous for its medical school (Bengtson, 1991).

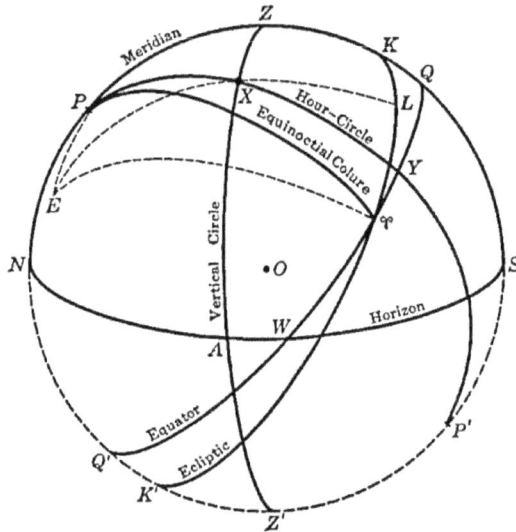

1a The ecliptic is the natural prototype of ellipsis, one of the three types of conic sections [Henry Norris Russell, Raymond Smith Dugan and John Quincy Stewart (1926). *Astronomy: A Revision of Young's Manual of Astronomy.* Vol. 1. Boston: Ginn].

The planning of the city of Alexandria was entrusted by Alexander, or more probably by Ptolemaios I the Savior, to Deinocrates of Rhodes, perhaps the most eminent architect of his time and a great city-builder. At the same place, Sostratos of Cnidus built a lighthouse on the little island Pharos, one of the Seven Wonders of the World (Sarton, 1959). Clearly, the Eastern Mediterranean, at least in antiquity, was outstandingly creative in the continuous quest for practical and theoretical innovation. In Alexandrian Renaissance, there was, in part, a convergence of civilizations.

During the Alexandrian Renaissance, there was a considerable theoretical progress in the adopted scheme of the world. There was no coincidence for the building of the Mouseion and the Library in Alexandria, because in Egypt one can find the longest, nearly straight, and wider areas for measurements, updated repeatedly by the 'bematists'. Based on similar traditions on measurement, Eratosthenes, perhaps the greatest scientist of the third century BC, estimated the circumference of the Earth by comparing the angle of the sun rays in Alexandria and Assuan (Fischer, 1975; Gulbekian, 1987).

The Alexandrian geographer Claudius Ptolemy (second century A.D.) completed the blank areas in the charts of his predecessors,

Strabo and Eratosthenes. Nevertheless, his map hindered all questioning about the validity of his worldview. Ptolemy was mistakenly representing the Indian Ocean as an inland sea; he was also filling up the Southern Hemisphere with Africa, ignoring Antarctica, altogether with America, of course. Regardless the mistakes, the value of Ptolemy's geography was coming from his idea of a unified global ocean. Later, in the fifteenth century, the Italian humanists translated the works of the ancient geographers, which influenced the explorers. Geographical conceptions were being gradually liberated from dogmatism, accepting the theory that the Earth is global and regenerating Ptolemy's belief that the European west coasts are close to the eastern Asia (Gaspar, 2013).

The next magnificent discovery revealed the genius of Diophantus, who is considered as the founder of algebra. Greek scientists as Diophantus introduced spatial terms in the methodological conceptions of theoretical mathematics. At first, the Greek term οδός, whose range of meanings includes "way," "general method," and "strategy." Secondly, the implementation of a method or strategy that begins with a separate stage which is discovery, investigation, in Greek εύρεσις, heuristics, or invention; nothing more than the establishment of an equation out of the problem, thus, only a primal part of the resolution (Christianidis, 2007). Thirdly, Diophantus acknowledges the Pythagorean terminology of numbers as 'squares' and 'cubes'; whereas, in his texts, the Pythagorean terms 'dynamodynamis', 'dynamokybos' and 'kybokybos' were also let in.

Space, knowledge, and power were grippingly linked in all the aforementioned scientific approaches. For this reason, the research includes challenging themes in review and literary critic about the practices of freedom, social relations, spatial distributions and space-language (Elden and Crampton, 2007) employed in the various disciplines that were integrated in the Alexandrian Library and Mouseion.

Chapter 2

Earlier than Western Science: knowledge in transcultural historical settings

The lack of empirical evidence was not always an obstacle in the quest for the wider possible overview upon the environment, in societies based on pre-philosophical and magical institutions. The Enuma Anu Enlil series offer an early important paradigm of theorizing based on omens and divination (Rochberg, 1999). "Shore-sighting birds" was another remarkable application, which proves the magnitude of the demand for rescinding the impediments and the distances in world history level.

This interest and want for concentration was even more aspirating in the field of economics, leading thus to the development of civilizations based on "large-scale societies whose members contribute taxes, labour, or tribute to the state and pay homage to their leaders" (Headrick, 2009: p. 17). The research on the congeries of large-scale global interactions and commodities-exchanges (what we call today Global or World History) was influenced by authors such as Richard Hennig (1936-39; 1944-1956) and Edward H. Schafer (1963). The latter made a listing of the Chinese foreign trade, including sheep and goats, camels, elephants, rhinoceroses, lions, parrots, peacocks, jewels, industrial minerals, gold and silver coins, books, tablatures, maps etc. Certainly, this shift in experience was increased at the break of the age of global trade, "when all manner of commodities – stones, woods, spices, herbs, metals and metallic wares – are eagerly sought for all over the earth; neither was mining carried on everywhere in early times as it is now", as William Gilbert (1893: p. 17) had stressed. For millennia, the main project of the Chinese State was the planning of water conservancy, irrigation and flood control programs, along the Yellow River firstly, by the Hsia and the Shang (Yin) dynasties, later to Yangtze River and across the country, with the proliferation of dikes, dams, canals, and artificial lakes (McClellan and Dorn, 2006). The Chinese produced very early hoes, iron plows, bronze bells, horsepower and breast-strap harnesses, double acting piston bellows, blast furnaces, crossbows, relief maps, calendars, paper, waterpower, canals, as the Grand Canal, which began in 486 BC during the Wu Dynasty; furthermore, rudders, masts and watertight compartments for shipping (Restivo, 2005).

The early period of the Chinese history of science is situated between 1520 BC and 221 BC, with hydraulic control, metallurgy, the one hundred schools, as Confucianism and Legalism in the front stage, altogether with Taoism, Mohism, the *Book of Changes* etc. Then, in the era of the New Dynasty of Wang Mang, the slavery was abolished, land was distributed, masterpieces were produced in the fields of astronomy, mathematics, and medicine. Simultaneously, works such as the *Santong Calendar* introduced the mathematical approach to the natural and social science, while "Wang Mang gauges (such as Lvjia Instrument) and measurement tools (such as Mang caliper)" (Liu, 2015: p. 22) gave evidence for the efforts to unify China.

> *Fine highways and waterways fostered overseas trade, but so did a change in the taste of the young sovereign Hsüan Tsung, who, at the beginning of his reign had an immense pile of precious metals, stones, and fabrics burned on the palace grounds to signalize his contempt for such expensive trifles. But a few years later, seduced by the tales of wealth from abroad accumulating in Canton, the emperor began to relish expensive imports, and to watch jealously over the condition of foreign trade. The old natural economy, under which pieces of taffeta were the normal measure of value and could be used for the purchase of anything from a camel to an acre of land... (Schafer, 1963: pp. 7-8).*

The year 731 AD was a turning point to the introduction of the money economy and foreign trade between the centres of Canton and Yang-chou, through the Korean territories and the Indian Ocean, to the remote harbour of Sīrāf in the Persian Gulf.

> *Chinese sources say that the largest ships engaged in this rich trade came from Ceylon. They were 200 feet*

long, and carried six or seven hundred men. Many of
them towed lifeboats, and were equipped with homing
pigeons. The dhows built in the Persian Gulf were
smaller, lateen-rigged, with their hulls built carvel-
fashion, that is, with the planks set edge to edge, not
nailed but sewed with coir, and waterproofed with
whale oil, or with the Chinese brea which sets like
black lacquer (Schafer, 1963: p. 13).

Much more dramatic, almost impossible without the Bactrian camel, was the overland caravan trade from the end of the Great Wall, through the Gobi Desert, Samarkand, Persia and Syria. In the late Song dynasty, gunpowder, the compass, navigation and the movable-type printing (replacing the woodblock printing on paper) were disseminated. The Yuan Dynasty attained great achievements in "astronomy, mathematics, agriculture, water conservancy, medical science, landscape painting, Zaju Opera, and Yuanqu Songs" (Liu, 2015: p. 24). Moreover, the most striking and impressive characteristic of the Chinese culture was the centuries-long coexistence with a series of different cultures. From the Sumer and the Akkad, at the times of the Banpo culture and the Erlitou culture in China, the Egyptian civilization, since 3000 BC, the Harappa culture of the Indus River Basin (2800-2000 BC), to the Aegean civilization, the Axial Period etc.

2.1. Reports on Chinese Science

In the field of education, the most traditional and marvellous invention of the Chinese was the abacus (by the second century BC), based on the decimal system. In the first century AD, the book *Nine Chapters on the Mathematical Art (Jiu Zhang Suan Shu)* included 246 practical measuring problems. During the fifth century, Tsu Ch'ung-chi (Zu Chougzhi) and his son found that $3.1415926 < \pi < 3.1415927$ and arrived at the rational approximation $\pi = 355/113$, which yields π correct to six decimal places. Then, in 1420, Jamshid Al-Kashr of Smirkand computed π correct to 16 places. Western mathematics did not surpass the approximation of Tsu until around 1600.

Chinese records of sunspots, comets, meteors, northern light and novae were available from four millennia. According to the oldest in the world known report on sunspots, in the third month of the year 28 BC "the sun at its rising presented a yellowing colour" and "a dark gas was observed in the midst of the sun" (Xiaozhong, 1989: p. 11). Chinese astronomers also wrote, over a period of more than

2,500 years, 31 reports on the appearances of Halley's Comet. The great Shen Kuo (1031-1095 AD), of the Northern Song Dynasty, mastered astronomy, calendar, music, medicine, physics, geology, mathematics and divination, and pointed out the phenomenon of magnetic declination.

Moreover, the Chinese calendar-keepers issued official calendars by the Xia Dynasty (twenty-first to seventeenth centuries BC) and developed the Astrometry and astrometric instruments, the armillary sphere, the Equatorial Torquetum, and the water-driven astronomical clock tower; they compiled catalogues of stars and star maps, they measured the meridian, and they advanced astronomical navigation. During the Northern Song Dynasty also, a water-driven astronomical clock-tower was built. Su Song (1020-1101 AD) constructed a water-powered armillary sphere and a celestial globe tower.

Before the first half of the Ming Dynasty, the Emperor prohibited any private occupations in Astronomy and astrology, because they ought to be exclusively governmental affairs. The oldest available collection of nautical maps and star charts (star bearing charts for crossing the oceans) was *the Zheng He Hang Hai Tu* (*Charts of Zheng He's Voyages*), reproduced in *Wu Zhi In* (*Treatise on Weaponry*) (Shuren, 1989; Ptak, 2007). The fleets of Zheng He carried magnetic compasses, star bearings instruments, charts and lead line. The Chinese geographers used the results of Zheng He's observations for the compilation of new maps, while Ma Huan presented his expeditions with Zheng He in the book *Yingyai Shenglan* (Parry, 1981). At the heart of it all, printing, gunpowder and the magnet were the most important mechanical discoveries, congregating greater power than any empire on earth, as Francis Bacon believed, without knowing that all of them had been Chinese inventions (Needham, 1986).

In general, the contribution of Chinese scientists is significant in the fields of medicine, papermaking technology, printing, gunpowder, south-pointing needle and compass, ceramics technology (pottery and porcelain making), weaving technology, astronomy, mathematics, physics, mechanics, optics, acoustics, metallurgy, botany, agriculture, biology, mining, mechanics, water, geoscience, construction, transportation and military technologies. Public works for the common welfare, irrigation, river control, water power, geomancy, pharmaceutical research, iron casting, encyclopaedic publication, and humanitarianism prove that science and technology in China is a cooperative enterprise.

2.2. Exchanges and diaspora

The knowledge of the author of the ancient mathematical text *Zhoubi Suanjing* (周髀算经) for the 5 frigid and tropic zones is currently recognised as surprising. In the *Zhoubi Suanjing* one may also find the Canopy-Heavens Universe Model, which complies with the model of the universe in the ancient Indian sacred scriptures book *Purāṇas* (Rocher, 1986), and was also present in the Introduction to Geography by Eratosthenes (275–195 BC).

In the Tang Dynasty, three Indian astronomers influenced Chinese science: Kasyapa, Kumara and Gautama, the most eminent among them. An offspring of Gautama translated in Chinese the Indian astronomical treatise *Navagraha*, which originated from Greek (Kropf, 2005). The Western Xia, the Western Liao and the Yuan-Mongolian Empire developed significant exchanges with the Islamic countries. Outstanding scientists were also Guo Shoujing[2] of the early Yuan Dynasty, the famous Persian astronomer Jamal al-Din, who directed both the Hui and the Han Bureau of Astronomy during the Juan Dynasty, and Xu Guangqi of the late Ming Dynasty. In the meantime, the cultural exchanges developed firstly between China and India, since the Han and Tang Dynasties, and then with the Islamic world, during the Song and Yuan Dynasties, and lately came the time of the juxtaposition with the Western civilization at the times of the Ming and Qing Dynasties.

At the end of the sixteenth century, the Jesuits visited China and the cultural exchanges between Far East and Europe were developed and extended (Jones, 2001; Mungello, 1989).

Missionaries of different nationalities, including Matteo Ricci, Ferdinand Verbiest, Johann Adam Schall von Bell, and Thomas Pereira, successfully entered the

---

[2] "He compiled, Calendar for Time Service, the ultimate calendar in the traditional Chinese calendars, supervised the manufacturing of quite a few instruments, such as the abridged armilla, the scaphe, the height meter, the shadow definer, the direction-determining board, and the ingenious planetarium, and he conducted unprecedented measurements for study of heaven" (Liu, 2015: 73).

palace. Their translation of Western scientific books to China shocked the Chinese literati in knowledge. Xu Guangqi, Li Zhizao, and other high-ranking intellectuals not only learned and accepted Western scientific knowledge, but also converted to Catholicism. During the reign of Emperor Kangxi in the Qing Dynasty, the French King Louis XIV, also sent a number of "King's mathematicians" to China. Some of them are scientists too. For example, Joachim Bouvet, Jean Francois Gerbillon later became Emperor Kangxi's teacher of Western mathematics, astronomy, and anatomy (Liu, 2015: p. 25).

Significant in this framework was the failure of the Chinese Directorate of the Astronomy, in 1629, to predict a solar eclipse with their traditional method, while Xu Guangqi made the right prediction following the Western method. Thereafter, the emperor Chongzhen assigned Xu to compile a new calendar. Xu Guangqi produced together with the Jesuit missionaries the *Chongzhen Imperial Almanac.*

2.3. The Islamic transfers of the traditional science

Many debates have arisen concerning the quality of the assimilation of foreign elements by the Islamic science, along with the economic or intrinsic causes of its decline, after a zenith around 1000 AD. The Islamic Empire originated from a particular blend of transnationalism, hegemonism, and localism, which quite early isolated the infidel populations. The intention to exclude any kind of strangers was obvious in the building of the new cities Kufa, Basra, Fustat, Marv and Qayrawan. Those cities were planned on the basis of imperviously congregated garrison settlements, for the purpose of avoiding the active interaction with foreign cultures (Donner, 1999). At the same time, aside from their invasion to the Iberian Peninsula (711), the Muslim rule expanded also to the East, from Bukhara and Samarkand to Indus River, between 705 and 713.

If we want to propose a fundamental *eastern* element that contributed to the building of modern science, then the Brahmi numerals may be a suitable candidate. The invention of the Brahmi numerals in India, by the third century BC, altogether with the instruction of the basic methods for calculating and the presentation of the algorithms by al-Khwarizmi, have played a critical role in the development of science, technology, industry, and commerce (Papadimitriou et al. 2010). In fact, Al-Khwarizmi, apart from the first Arabian treatise on algebra, wrote a small book upon the calculus of India. The Latin translation of that book disseminated in Europe the knowledge of the Hindu-Arabic numerals.

One of the results of daring transnational contacts was the implementation of ambitious translation projects that presupposed the excellent knowledge of the Greek, Syriac, Persian and Indian languages. During the transition to the period of assimilation of the traditional sciences, al-Faz r translated, in 154/770, the astronomical *Siddhānta*. The adaptation was obvious in the task of understanding the use of the astrolabe and the armillary sphere.

An early Arabic translation of the Greek alchemical treatise of Zosimos dates from 38/658. Other important translations were the medical textbook (*Kunnāš*) of the Alexandrian presbyter Ahron, the astrological "Book of Fruit" (καρπός; *Kitāb aṭ-Ṭamara*), the *Zīğ aššahriyār* (an editorial of the Ptolemaic Canon on the basis of Indian tables), Euclid's *Elements*, several works of Archimedes, and many of Aristotle, e.g. some parts of the *Organon* (from Middle Persian translations), and a translation of *Kalīla wa-Dimna*. The Arabs made also significant progress in the areas of literature and literacy, according to Fuat Sezgin (2003).

The advances were remarkable in the field of astronomy, with the translations of the Ptolemaic *Mathematical Syntaxis* (renamed *Almagest*) by Yahya ibn Khalid al-Barmak, the astronomical works of al-Khwarizmi, Thabit ibn Qurra, al-Battani, al-Farghani, Abd-al Rahman al-Sufi und Ibn Yunus, and the *al-Zij al-Mumtahan*, a set of astronomical observations for the verification and the correction of the Ptolemaic tables. The mathematician Habash al-Hasib supplemented the Ptolemaic astronomical works by introducing primitive trigonometric functions of sine, cosine, and tangent with his work *al-Zij al-Dimashqi* (Dallal, 1999).

The library Bayt al-Hikma gathered systematically, since 832 AD, translations of scientific works directly useful to the caliph, especially in the fields of medicine, astronomy, astrology, alchemy, and logic. For instance, the most famous among the translators, the Nestorian Arab Hunayn ibn Ishaq al-Ibadi (808–73), translated many works of Galen and Hippocrates, for the governor of Bagdad.

The founder of the House of Wisdom (the Bayt al-Hikma), Caliph al-Ma'mūn (r. 813-833), organized also the translation of Greek scientific manuscripts from Byzantine sources (such as Ptolemy's

Γεωγραφική Υφήγησις) and commanded astronomical observations, the determination of the difference in longitude, the measurement of the earth's surface, the establishment of observatories in Damascus and Baghdad (Sezgin, 2003). "The best known, established in 1259, was the observatory at Maraghah in a fertile region near the Caspian Sea" (McClellan and Dorn, 2006: p. 110).

From the *Kitab al-Jabr wal-Muqabala* of al-Khwarizmi (ca. 780-850) to the works of al-Birouni (973-1048), the influential potentiality of the Islamic science became prominent. However, it would be impossible without the transfer of foreign knowledge. The Arab physicist Ibn al-Haytham (c. 965–1039 AD) presented an impressive multi-volume work in optics, with contributions in the subject of dioptrics, mainly about the refraction of light. Ibn al-Haytham followed the concept of Ptolemy's Optics and incorporated in his analysis of reflection the valuable knowledge of Heron and Diocles. After the ascendance of Kublai (the first emperor of the Yuan Dynasty) to the Chinese throne, the Persian astronomer Jamal al-Din translated the names of seven astronomical instruments (armillary sphere, organon parallacticon, tellurium, astrolabe etc.) and described their structures and applications (Jiang, 2015).

In the mid-thirteenth century, the Mongols - who learned the gunpowder from the Chinese and used Chinese artillerymen in their armies - transferred it to Islam. The Chinese used the saltpetre widely from the Han Dynasty, described the production of the gunpowder in the nineteenth century military treatise Wu Ching Tsung Yao (武经总要) and developed from the tenth century firelance, explosives, grenades, bombs, fire arrows, cannons, mobile rocket launchers, landmines, watermines etc., with the use of sulphur, saltpetre and charcoal (Goodrich and Chia-Shêng, 1946; Ling, 1947; Khan, 1996). The Arabs named the saltpetre *Chinese snow* (thalj sīnī) and the so-called Military or Gunpowder Revolution became considerable from Europe and South Asia, to Korea and South-East Asia. In 1293, a Yuan fleet carried Chinese bombards during its invasion in Java (Di Cosmo, 2002; Collier, 2004; Cook, 1994; Hodgson, 1974).

2.4. The Transition to the Modern Era

Abdallāh Ibn Baṭṭūṭa (1304-1369) was a great world voyager who travelled from his hometown Tangiers to Alexandria, Cairo, Syene, Syria, Mecca, East Africa, Mozambique, Asia Minor, Byzantium, southern Russia, Central Asia, India, the Malay Peninsula and China (Dunn, 2012). Ibn Baṭṭūṭa's case, through his globalising viewpoint, offers a linkage between the Middle Age and the Renaissance.

Moreover, the slavery and its abolition, as in the cases of Indian Ocean, Africa and America (Allen, 2010; Bose, 2006; Campbell, 2005), reveal important issues for the historical studies of the diffusion of technology and global trade. The slave trading activities of the Vāniyā caste Gujarati Indian merchants (Machado, 2005), for instance, constitute a significant part of the transoceanic commercial networks of the global Hindu Sindhis from Bukhara to Panama (Markovits, 2000).

Forms of resistance to the modernization appeared worldwide during the dissemination and the transfer of industrial technologies. Chinese population cancelled, in June 1865, the construction of telegraph lines by a British man in Shanghai, on the ground of their possible incompatibility with the traditional geomancy belief system of feng shui (風水). The same year, the Chinese Gendarmerie dismantled a railroad built by a British merchant in Beijing (Yongming, 2006).

The introduction of technical infrastructures, such as transport and communication, was excessively influential to the creation of powerful transnational networks across borders. Economic organizations, transportation companies, trade firms, travel agencies and high technology multinationals became gradually the leading agents of a transnational transformation of the globe. The formation of a global world means also that warfare and religion lost their central, dominant role, whereas the technological transfers, after 1500, and especially the print capitalism (Anderson, 2006), multiplied the ecumenical attitudes in every different culture. However, the most powerful factors of modernization, as Appadurai (2005) supports, are the electronic media and the migration, with their joint, ponderous impact in the field of the quotidian imagination of the ordinary people.

The transnational aspect of historiography has found fertile ground in the Americas, among the workers, the immigrants, the artists and the litterateurs of the new world, no matter if they were

European, Asian or African in origin. In fact, there are two main lines of interaction here, the transatlantic and the transpacific respectively (Fink, 2011; Liu, 2005). Nevertheless, at the turn of the century, the anthropological studies made significant progress with the works of scholars such as Jeremiah Curtin (1996), mainly with his study of the unification process throughout the Silk Road under the Mongolic imperium (Gladney, 1999).

The debate about the British critical contribution to free trade, imperialism, and global history refreshed in the 1950-60s with the Robinson-Gallagher controversy (Gallagher and Robinson, 1953; Robinson et al. 1961). Thirty years later Cain and Hopkins (2001) introduced an argument, which focused on the role of the City elite to the formulation of British imperialism.

At the same period of time, the social studies of science and technology accomplished an interesting passage from the Actor-Network Theory (Latour, 2005) to the social theory of the market construction, connecting thus a sociology of translation, with ethnography, economic geography, and epistemology (Ouma, 2012). Nowadays, the transnational character is present in diaspora-communities such as that of the Ismaili movement during the last one thousand years, from the Houses of Wisdom in Bagdad and Cairo to the initiatives of the Aga Khan Development Network in Tanzania (Kaiser, 1996)

Chapter 3

The age of the oceanic discoveries

During the Renaissance, humanists translated the works of the an-cient geographers, which shaped the worldview of the first great explorers. Geographical conceptions were gradually liberated from reactional taboos, accepting the theory that the Earth is global and regenerating the belief that the West European coasts are closer to the Eastern Asia. The geographers were envisioning the idea of a unified global ocean and the explorers discovered sea passages, oceanic spac-es, and zones of continuous winds and currents.

Christopher Columbus in 1492 discovered the Bahamas, Cuba, and Haiti, and later the Antilles, the Caribbean coast of Venezuela and Central America. The travels of Vasco da Gama and the circum-navigation of the globe by Ferdinand de Magellan are considered as pivotal for the discovery of the naval route to India, and also for the invention of America and the Pacific Ocean. Subsequently, the borders of the ancient *Orbis Terrarum* were being erased.

The Portuguese overseas Empire

The first recorded Portuguese ocean voyage was probably carried out in 1341 to the Canary Isles, the so-called Fortunate Islands by the ancients. The epoch of the Portuguese overseas empire began in the early fifteenth century, with the fall of Ceuta to the hands of Prince Henry the Navigator that was followed by voyages to the Canaries, the Cape Non and the Cape Bojador.

> *According to Diogo Gomes, D. John de Castro made his attempt on the Canaries in 1415 and not in 1425, and as he reported the existence of a strong current be-tween the islands, Henry sent out Gonçalo Velho in 1416 to find out the reason of it, the first scientific expedition of this kind recorded* (Prestage, 1933: p. 28).

The plans of Prince Henry, according to the chronicler Zurara, were to discover lands beyond the Canaries and Cape Non, such as Guinea, to establish trading relations and ascertain the Christian boundaries with Islamic lands. Until the late 1440s, slave raiding and slave trade became the main objectives in Prince Henry's expeditions. Meanwhile, the Portuguese organized the colonization of Madeira in the 1420s and the Azores in the 1440s. From 1448 they engaged also in gold dust trading (Disney, 2010). Aside from many doctors and astrologists, Prince Henry hired Master Jacome, an expert cartographer and maker of nautical instruments, to assist him in cosmography. Master Jacome has been identified with Jahuda Cresques; whose father Abraham produced the Catalan Atlas in 1375.

Furthermore, Henricus Martellus was the cartographer of the voyages of Diogo Cão and Bartolomeu Dias, which resulted in the discovery of the southern Cape of Good Hope. In 1498, after Cabral's and Pacheco's excursions to Brazil, Vasco da Gama, sailed over the Cape of Good Hope to arrive at Calicut. The earliest surviving map of the eastern and western voyages of Cabral, Da Gama and their precursors is the world map of Alberto Cantino (1502), which depicted, among others, the lately reached Americas and the Tordesillas line.

The Red Sea and the Persian Gulf remained for centuries the main oceanic routes to the Far East. From the very beginning, Vasco da Gama "carried out a heavy bombardment of the town of Calicut, an important event in the history of naval gunnery as well as in that of Indo-European relations" (Parry, 1961: p. 39), while Afonso d'Albuquerque's task was to destroy the Arab spice trade by the use of guns. As a second Viceroy, Alfonso de Albuquerque conquered Ormuz in 1508, Goa in 1510, which became the capital of his colonial empire, and Malacca in 1511, which came to be a commercial and diplomatic center. The Portuguese explored the Spice Islands in 1511-12, whereas Francisco Rodrigues produced maps of the China Sea, influenced by local cartographers.

In Portugal, the administration responsible for overseas territories was the *Casa da Mina e India*, which constructed model charts (such as the *Padrão Real*) and instruments, developed navigational knowledge, examined and collected information from the pilots (Barrera-Osorio, 2006). The training and the certification of the pilots was organized by a Cosmógrafo-Mor, i.e. chief cosmographer, with expertise in mathematics, cosmography, and astronomy.

Between 1540 and 1759 the Society of Jesus developed influential scientific activities in Portugal. "In 1574, following a demand made by the king, the Jesuits created their first mathematical class in Portugal—the Aula da Esfera (Course on the Sphere)—at the Colégio de Santo Antão, in Lisbon" (Fontes da Costa and Leitão, 2009: p. 42). That center provided technical and mathematical training to the navy.

Cosmographers and pilots

Charts, ships, guns and navigation tools were the most general categories of the tools used by the great oceanic explorers. From the thirteenth century, at least, Catalan and Italian hydrographers drew *portolani*-charts, based on practical knowledge. Great cartographers of the renaissance were Bartolomeo Pareto, Battista Beccario, Zuane Pizzigano, Martin Waldseemüller, Andrea Bianco, Grazioso Benincasa, Juan de la Cosa (Columbus's pilot), Luis Teixeira, Gabriel de Valseca, Matteo Ricci, Diogo Ribeiro, Abraham Ortelius, Juan López de Velasco and many others (Marques, 1995).

> The maps of Andrea Bianco (Bianchi), along with those of Walsperger (1448), the Catalan-Estense map of 1450, the Borgia map, the Genoese map of 1457, and Fra Mauro's map of 1459 (...), form the beginnings of a transition period, away from the circular, Jerusalem-centered religious depictions of the earlier medieval mappaemundi, and toward those that were to form the Renaissance period of cartography.[3]

In 1445, the Portuguese Dinis Dias discovered the mouth of the River Senegal and Cape Verde. The west coast of Africa was depicted in the chart of the Venetian Andrea Bianco, in 1448. Twenty years later (1468), Grazioso Benincasa of Ancona drew the chart of the discoveries of Gambia, Rio Grande, Cape Verde Islands and Sierra Leone. By the Gulf of Guinea, while reaching the Equator, the sea-

---

[3] Bianco World Maps, http://cartographic-images.net/ Cartographic_Images/241_Bianco_World_Maps.html

farers were disappointed by finding the coast leading far south (Skelton, 1958: p. 28). Around 1462, European sailors had managed to calculate the latitude from the altitude of the Pole Star.

After 1485, the navigators used tables of the sun declination for the observation of latitudes in the southern hemisphere. Dead reckoning, based on the combination of charts, log crude speedometer reading, and stars, was indispensable for the first explorers. Sandglasses, pocket watches, compasses, and gimbals were also used for the discoveries, all along with the astrolabe, the quadrant, the armillary sphere, the lead line and the chip-log, which was introduced in the early sixteenth century.

> *The Portuguese explorers of the African coast pre-*
> *ferred, whenever possible, to take their sights ashore.*
> *They stood in towards the coast, anchored, pulled*
> *ashore, and hung their astrolabes from tripods set up*
> *on the beach. From this position they took their noon*
> *sights and worked out their latitudes with, on the*
> *whole, surprising accuracy* (Parry, 1961: p. 20).

The earliest signed Portuguese nautical chart was constructed circa 1485 by Pedro Reinel, based on the explorations made by Diogo Cão in Central- and South-West Africa (1482-1485). Pedro Reinel was also the first cartographer who presented, in 1502, a chart with "a meridian line graduated in degrees of latitude" (Skelton, 1958: 35). The pilots used "the Jacob's staff for measuring the elevation of the North Star, and the astrolabe for determining the height of the sun at noon" (Portuondo, 2009a). In 1503 the Spanish Crown assigned the construction of precise transatlantic charts and astronomical instruments, and the training of the pilots, to the *Casa de la Contratación* (House of Trade) in Seville. Later, in 1524, the *Consejo de Indias* (Council of Indies) and the *Real Corte* were established for the regulation of affairs in the Spanish territories. Famous cosmographers were employed in those organizations, while their continental colleagues, such as Gemma Frisius, Sebastian Münster, André Thevet and Peter Apian, incorporated their findings in the general cosmographies.

The works of the Spanish cosmographers, such as the *Arte de navegar* (1545) of Pedro de Medina and the *Breve compendio de la sphere y de la arte de navegar* (1551) by Martin Cortés, were published in many other countries and languages. In 1582, Philip II appointed the cosmographer Jaime Juan to teach the pilots how to use the navigational instruments, to make maps and determine the latitude and the longitude.

The technological skills of scientists such as Christóbal Gudiel, the questionnaires upon the geographical expositions (Relaciónes), the magnificent botanical surveys in the New World and the creation of a mathematical academy in Madrid were further significant issues (Goodman, 2009). The formal education of the cosmographers and the personal experience of the pilots were essential for the implementation of the plans of the European Kingdoms over transoceanic routes. The need to introduce the Portuguese navigational astronomy in Spain created the office of the "piloto mayor" (Navarro Brotons, 2000). Amerigo Vespucci was the first "piloto mayor" (1508-12) who interviewed the pilots returning from the Indies. Sebastian Cabot served in the same position from 1518 to 1548, when he went back to England.

3.1. Sea passages

The westward passage to Asia was one of the study cases that occupied Roger Bacon in his *Opus Majus* and Cardinal Pierre d'Ailly in his *Imago Mundi*, an essential guide for the explorers. The development of cartography stimulated the debates about sea passages. In 1459, the cartographer Fra Mauro produced the most precise and exhaustive map of that period, the first map that mentioned "Zimpagu" (Japan). At the same period, the proposals for Northern passages referred to the route 'toward the Orient', 'towards the Occident' and 'right toward the Pole Antarctike'.

The seamen and merchants of Bristol were trading regularly with Iceland in the fifteenth century, where they heard of Greenland, Markland and Wineland the Good. Between Bristol, Azores and Madeira there were also significant commercial relations. John Cabot singled out Bristol as starting point for his voyage to Nova Scotia, in 1497. After his first journey, Cabot mistakenly reported to the King Henry VII that he had reached Asia. In his second voyage, Cabot's mission disappeared. In 1498, the King Manuel of Portugal sent a mission to investigate what John Cabot could be supposed to have found, probably, in the fisheries and the timber around Grand Banks of Newfoundland. Next year, after Vasco da Gama's return from India, the Portuguese Gaspar Corte Real sailed from Lisbon to Greenland, where he was stopped by ice. In 1501, Corte Real's mission made another attempt to discover a polar passage between

Greenland and Labrador, where they were lost. Gaspar's brother, Miguel Corte Real also disappeared on the same route, in 1502. Sebastian, John Cabot's son, managed to pass the Hudson's Strait and entered the Hudson Bay in 1509, where his men insisted on returning back to Bristol.

On the first printed world map, authored by G.M. Contarini (1506), an outspreading sea passage divided the North American lands discovered by the English and Portuguese from the Spanish discoveries in South America. However, the real sea passage was represented in a chart by Antonio Pigafetta, who sailed with *Victoria,* one of Magellan's ships. Moreover, Sebastian Münster in 1540 designed a map of America and the Far East.

The *Pilot Mayor* Amerigo Vespucci should be recognised, according to one of his letters, as the explorer who consciously discovered the new continent. In 1507, the German cartographer Martin Waldseemüller reprinted Vespucci's letter and produced a world map that represented the new lands, naming the southern of them 'America'. The map, altogether with the related globe and the textbook *Cosmographiae Introductio,* were included in the humanist activities of the St-Dié group at the Gymnasium Vosagense in Lorraine (Johnson, 2006).

The worldview changed rapidly in a few decades. The cartographer Piri Reis compiled a world map from around twenty charts and Mappae Mundi produced by Columbus's cartographers (Tebel, 2012). The invention of the mechanical movable type printing in the middle of the fifteenth century had a critical influence to the dissemination of the cosmographical knowledge, although the Portuguese and the Spanish Crown, especially Philip II, prohibited the open access to the cartographical works of their cosmographers (Portuondo, 2009b).

From the end of the sixteenth to the beginning of the seventeenth century the Northeast, the Southeast and the Northwest passages to the Pacific were some of the most challenging goals of exploration. Martin Frobisher, Henry Hudson, the Danish Arctic expeditions tried also to explore the Northwest Passage. Another proposal was the circumnavigation of America from the Strait of Magellan, to the Pacific and thence to the Strait of Anian (Behring Strait). In 1577, John Frampton translated the Marco Book in English. In 1569, Mercator presented a world chart and in 1580 he conjectured a Northeast Passage to Cathay (Northern China).

Until the English journeys, in 1555-57, to the south of Novaya Zemlya, the North-East Passage was known only as far as Vardö, and it was believed to have been navigated to the mouth of the River Ob. Further east, were placed 'Cape Tabin' and the so-called Strait of Anian (between Asia and America). From 1555 to 1564, trade relations between Russia and the explorers' Muscovy Company were established. In 1594, Dutch expeditions were sent to explore the passage to the north of Novaya Zemlya, but they preferred the southern passage through Yugorsky Strait.

3.2. Ships and shipbuilding

The European ships, from the beginning of the fifteenth century to the end of the sixteenth, surpassed the Chinese junks and became the best in the world. The galleys, although very common in trade until the eighteenth century, they were not suitable for exploration. The shipyards built, since 1400, heavy, stable, square-rigged ships, with topsails and castles fore and aft for cross-bowmen. The Portuguese, however, preferred the lateen caravel, similar to Arabian ships such as the *baghlas* of Persian Gulf and the *Sambuks* of Red Sea.

"The lateen sail is the special contribution of the Arabs to the development of the world's shipping; it is as characteristic of Islam as the crescent itself. It is also a very efficient general-purpose rig. The qualities of any sail when beating to windward depend largely on its having the leading edge as long and as taut as possible", as Parry (1961: p. 23) emphasized. The caravels were improved gradually, based on their behavior in exploration. The yards were shortened, set more nearly upright, fitted better to the masts; and a third, mizzen mast was added.

In the end of the fifteenth century, the *caravela redonda* combined a square-rigged course and topsails on the foremast and lateen-rig on main and mizzen. From the middle of the fifteenth century the fighting ships carried guns, usually brass artillery, mounted in the castle structures fore and aft, substitute for crossbow and arquebus fire. In the Portuguese caravels, which were lacking forecastle and fighting tops, the guns were mounted in the bows, on the poop, and across the gunwale. From the late fifteenth to the early sixteenth century broadside fire was further developed with embrasures and hinged scuttles.

3a.The naval battle of Lepanto. Antonio Lafreri: *Lordine tenuto dall'*
armata della Santa Lega Christiana contra il Turco **(Royal Geographical**
Society, London, 264.G.2)

The development of the *caravela redonda* was the result of the
combination of lateen and rig sail. However, from the 1530s, the
competition between the Iberian and the English and French fleets,
led to the construction of bigger ships (Fuchsluger, 2013). The naus
were developed by the Portuguese and the Spanish. They were
bigger and heavier than the caravels and the cogs, with two or three
masts. The ships used in the *carreira da India* were mainly galleons
and carracks (naus). The carracks were large merchant ships, while
the galleons were primarily fighting-ships.

Although many times the naus and the carracks are considered as
same, the size and the castles were the essential distinctive features
between them. The carracks were generally 40 to 50 meters long
and capable of carrying 500 tons. Their four-mast rigging enabled
the movement of exceptional freights and large caliber cannons
(Marboe, 2009). Columbus' fleet of 1492 comprised two small trading caravels, Niña and Pinta, and the nau Santa Maria, a ship-rigged
cargo carrier. Dias' discovery of the Cape of Good Hope, had been
accomplished with small mercantile caravels. The fleet of Vasco da
Gama was dressed up by that of Bartolomeu Dias, whereas two of
the ships, the flagship San Gabriel and the San Raphael, were constructed and equipped especially for the voyage. Magellan's cir-

cumnavigation of the world (1519-22) was achieved with the nau Victoria, while Captain Cook used a northern-seas collier barque. The Portuguese recognized the superiority of the Indian teak and constructed their *carreira* carracks at Goa, Cochin, and Bassein. The crews, especially in the Indian Ocean, were not European but Muslim, as Vasco da Gama's pilot Ibn Majid. Many times the sole exception to the non-Portuguese crew was the captain, who used the *roteiros* (Boxer, 1959).

Shipwrecks and representation

The ships are historically considered as the *most complex artefacts* produced for millennia at a global level. The evidence of ships in maritime archaeology refers not only to shipwrecks but also to offloaded ballast or jettisoned cargos (Conlin and Murphy, 2002). Most of the wrecked carracks were overladen, inefficiently stowed and belated. The enormous shipwreck-rate led also to harsh punishments, such as the hanging of the officers of the galleons *Santo Milagre* and *São Lourenço*, wrecked in 1647 and 1649 respectively. The Cape of Good Hope was known as the Cape of Storms, a site of numerous wrecks. In the *S. Paulo* narrative, a sermon delivered by Father Pedro Manuel Álvares, refers to the shipwrecks of the galleon *São João Baptista* and the *São Bento* that wrecked off the Cape of Good Hope, as instances of disasters caused by discord. The hopeful message is transmitted: "with concord *[concórdia]* small things grow into great, whereas with discord *[discórdia]* great things decline and diminish" (Blackmore, 2002: pp. 35-36).

Between 1497 and 1650, more than twenty-five percent of the 219 Portuguese shipwrecks were lost in the Mozambique Channel. Madagascar is a notorious shipwreck island, where castaways from the Indian Companies, and also pirates had suffered. In 1506, *São Vincente* was one of the first Portuguese ships that wrecked in their attempt to explore Madagascar (Van den Boogaerde, 2010). The extremely long and dangerous passage off South African capes and the seasonal character of the monsoons generated as a necessity the East India Companies, with their shelters around the Indian Ocean. The voyages from Lisbon usually started at the end of February or in March. They sailed from Goa back to Portugal at the end of December. The most shipwrecks occurred on the homeward trip before the carracks pass the Cape of Good Hope (Boxer, 1959).

> *In the eighty odd years from Vasco da Gama's first*
> *voyage to the union of the Spanish and Portuguese*
> *crowns, 620 ships left Portugal for India. Of these, 256*
> *remained in the East, 325 returned safely to Portugal*
> *and 39 were lost. In the next forty odd years - from*
> *1580 to 1612 -186 ships sailed, 29 remained in the*
> *East, 100 returned safely, 57 were lost. In the first pe-*
> *riod, therefore, 93 per cent of the ships which sailed*
> *from Portugal reached their destination safely; in the*
> *second period only 69 per cent found harbour* (Parry,
> 1961: p. 95)

A very common artifact found in many Spanish shipwrecks was the olive jar. The earliest European objects discovered in the New World were "a solid-gold crucifix, a gold bar, silver disks, cannons and three extremely rare, sixteenth-century astrolabes" (Orser, 2002), removed from the wreck of the *Espiritu Santo* in Padre Island, Texas, in 1967. Perilous sites, such as Goodwin Sands and Blackpool in England, Dry Tortugas in Florida or Yassi Ada in Turkey, and anchorages are regarded as ships' graveyards. Other shipwreck sites relate to particular activities, such as Red Bay in Labrador relates to the early Basque whaling industry (O'Leary, 1997).

3.3. Colonialist competition and utilitarianism

The maritime revolution was not a linear increment but a controversial process. In the eve of the Discovery Age, a sharp confrontation developed between the 'Silver Empires,' on the one side, and the resistance on the other, as expressed by pirates, privateers and "the *Cimarrones*, or Maroons, a tribe of runaway Negro slaves and Indian women who lived in the jungles of the Isthmus and defied all Spanish attempts to bring them under control" (Silverberg, 1997: p. 249). Saying this, one should stress that the linkages between reconnaissance, maritime and global history are systemic, that is to say, they relate to conquest, power, economic interests, piracy and the respective legal arguments, such as the *Mare Liberum*. For instance, the rise of Charles's V imperium, in 1519, was simultaneous with Hernán Cortés's military campaign to America and with Fernão de Magalhães's expedition in search for an access to the Moluccas from the east.

Moreover, the duel for European supremacy between Francis I and Charles V began in 1521. After 1564, two armed fleets of "twenty to sixty sail, usually escorted by from two to six warships" (Parry, 1961: p. 74) carried the bullion cargoes to Spain, while no other ship was allowed to cross the Atlantic, except from those two convoys. In the 1550s the amalgamation process, a more efficient method to extract silver, was developed in New Spain and helped to increase the amount of bullion exported to Spain, and the rest of Europe (Barrera-Osorio, 2006).

The circumnavigations of the world by Ferdinand Magellan and Francis Drake became a famous theme of the early modern cartography. The world's first circumnavigator was Ferdinand Magellan, with his ship Victoria. In 1505, Magellan had followed the Almeida expedition, which established Portuguese coastal fortresses in Sofala, Kilwa, Anjediva and Cannanore over the Indian Ocean. The capturing of the spice trade and the establishment of a Portuguese Viceroy required a handful of fierce battles in the Indian Ocean. The dramatic conflict was inflamed by Venice in coalition with Egyptian and Indian opponents of the Portuguese expansion.

Improvements of great value for the geographical science became manifest with the world map of Martin Waldseemüller, Magellan's exploration (1519), the cartographic technique of projections, developed by Gerard Mercator (Taylor, 2004), the voyages to Philippines by Miguel López de Legazpi and to Acapulco by Andrés de Urdaneta (1565), and the maps of the Pacific Ocean, produced by Diogo Ribeiro in 1529 and Matteo Ricci in 1584. Those maps created discussions about the repositioning of trade to the direction of the *South Sea*. The voyage of Álvaro de Mendaña to the Solomons (1567–68) was a step for the discovery of Oceania, while the circumnavigations of Francis Drake in 1577, Thomas Cavendish in 1586–88 and Olivier van Noort in 1598–1601, were mainly hunting for *bouillon* from the Spanish galleons (Camino, 2005).

The instability and the hostility increasingly dominated, as the Treaty of Cateau-Cambrésis did not terminate the French piracy against the Iberian fleets, while the rise of the Protestants in England and the Calvinists in Scotland coincided with the opening of piratical and slaving enterprises by English privateers such as Hawkins, Frobisher and Drake. In 1569-1572, the alliance between Dutch, Huguenots and English privateers cut the communications between Spain and the Netherlands.

Hoornse Eijlandt
Isle de Hoorn.

3b Illustrations de Journal ou description du merveilleux voyage (...) fait des années 1615, 1616 et 1617. Les marins de l'équipage de Guillaume Schouten repoussent à coup de fusil les Indiens de l'île de Hoorn. Neither the Straits of Magellan nor the Cape of Horn were used for ordinary trading lines, because the Pacific maritime commerce was organized by the viceroy of Mexico. The regular maritime connection between Manila and Acapulco in 1571 has been stressed as a hallmark for the creation of a global trade network and a coherent world market (Flynn and Giráldez, 1995; Vogl, 2013).

The conflict between England and Spain turned to open war, whereas Francis Drake was "singeing the King of Spain's Beard" by raiding and burning Spanish ships in Cádiz and Lisbon. The following year, in the autumn of 1588, around half of the ships of the Armada were sunk in the storms that raged around Scotland and Ireland. The tragic element, once more, meant to be uncovered not only with the shipwrecks but also by many violent British assaults against the French-American Huguenots, in the frames of reconciliation with Spain, before and after the English-Spanish War. At the same period, the Dutch West India Company was also organizing piracy and colonies from the Caribbean to Canada.

Based on the respective imperialistic competition, the notorious Black Legend was constructed by the works of William of Orange, Jan Huygen van Linschoten, Theodore de Bry, while Bartolome de las Casas, Alonso de Ercilla y Zúñiga, etc. criticized Spanish colonialism. The theological and the mythical element in the narratives, as it was shown in the *Libro de las Profecias of Christopher Columbus*, was further stressed by the absence of early modern ship technical drawings and maritime archaeological evidence (West and Kling, 1991).

On the contrary, the utilitarian option of opinion was expressed in studies such as Gonzalo Fernández de Oviedo's *Historia general y natural de las Indias Occidentales* (1535). Gold and silver, pearls, diamonds, amber, musk, tapestries, ebony, calico, cloves, pepper, cinnamon, mace and nutmeg were some of the precious commodities of the oceanic trade, as with the prey of the English pirates over the Portuguese galleon *Madre de Deus* in 1592 in Azores (Sobel, 1995). Very important, pepper, mainly from south-western India, prevailed in the spice trade; furthermore, ginger, saffron, rhubarb, grown in China, precious stones, emeralds from India, rubies from Burma, sapphires from Ceylon etc. (Fernández-Armesto, 2009).

> *It was in 1503 - six years after Vasco da Gama rounded the Cape of Good Hope - that the Portuguese sent their first spice-fleet to Antwerp... The term 'spice' then covered all manner of oriental luxury products - pepper, cinnamon, mace, nutmegs, cloves, pimento and ginger (used medicinally), together with sandalwood (employed as an astringent and blood-purifier), spikenard, the oriental gum-resin known as galbanum (much appreciated by women), wormwood, ambergris, camphor, ivory and various other rare commodities, all valuable and some hitherto unknown in Europe* (Roth, 1977: p. 21).

Treasures, such as ambergris, were sought in Madagascar, cacao and *xocolatl* were imported from South America. A Dutch cargo to Batavia around 1735 could contain wood, building bricks, iron, gunpowder, and wine, as well as chests with gold and silver ducats (Missiaen et al. 2012).

The rise of the British was based on piracy, slave trade and intensive slave labor in sugar colonies of the plantation complex, e.g. Brazil, Jamaica, Barbados and Saint Domingue (Curtin 1990). The Seven Years War (1756-1763), the treaty of Paris in 1763, the British control over the Bengal gunpowder production, the increase in industrial production, the establishment of the Bombay shipyard in about 1675 for the production of ships of Indian teak,[4] related to the rise of the British naval power and the emergence of the Lloyds underwriters.

3.4. Scientific discoveries, telescopes, clocks and longitude

The interaction between scholars and artisans, during the commercial and imperial expansion, was vital for the scientific advance in the sixteenth and seventeenth centuries. The mathematical and technical education, in the reign of Philip II, included the training of the engineers, architects, pilots, cosmographers, gunners and other specialties. Astronomical knowledge, outstandingly represented in the sixteenth-century Spain by Jerónimo Muñoz, was officially considered as necessary for various disciplines and tasks, that is, astrology, reformation of the calendar, geography, cartography and navigation (Navarro Brotons, 1992).

The scientific revolution started early in the 1520s, with the development of empirical practices, when the explorers confronted unfamiliar natural entities in the New World. For instance "a tree called bálsamo in Spanish and boni, guacunax, or canaguey in the native language, depending on the province" (Barrera-Osorio, 2006: p. 16). Furthermore, the medical and astronomical books, e.g. the astronomical Alfonsine tables, were preserved in the University of Alcala, while new tables such as the "ha-Ḥibbur ha-Gadol" were composed in the University of Salamanca (Feingold and Navarro Brotons, 2006).

The problem of longitude, "the height of east and west," was addressed by Galileo, who used the telescope for celestial observations and suggested that the observation of the longitude could be effectively accomplished through timetables of the disappearances and reappearances of Jupiter's satellites. However, the court of King Philip III of Spain rejected his proposal. Giovanni Domenico Cassini, a professor of astronomy at the University of Bologna pub-

---

[4] Teak, the commonest shipbuilding timber in the Indian Ocean, is an oily wood which preserves iron, unlike oak, which corrodes it; teak-built ships, therefore, are not subject to iron-sickness, as oak ships are (Parry, 1963).

lished, in 1668, the best set of astronomical observations. Cassini was appointed by Louis XIV, as director of the newly founded astronomical observatory of Paris (Sobel, 1995).

Christiaan Huygens constructed the first pendulum clock in 1656, declaring in his treatise *Horologium* (1658) that his clock was an instrument capable of establishing longitude at sea (Howard, 2008). The following years Huygens tested his clocks in shipping conditions. In 1664 he published the *Kort Onderwijs*, a manual for marine timekeepers.

By 1675, seeking for timekeeping stability in stormy ocean waves, Huygens presented the *spiral balance spring* that offered an alternative to the pendulum. This invention caused bitter strives between Huygens and Robert Hooke, regarding the patent of the spring balance watch. After the horrible Skilly naval disaster of 1707, the British Parliament offered the Longitude Prize. For this purpose, from 1730 to 1776, the carpenter and clockmaker John Harrison constructed various marine chronometers, awarded by the Board of Longitude.

The magnetic compass, mounted on gimbals and enclosed in a binnacle, had been proposed as a convenient instrument for observing longitude, because of the difference between the magnetic and the celestial North Pole according to longitude. The magnetic North Pole overlaps the actual pole in the Pacific while differing in the mid-Atlantic Ocean. The magnetic variation and the convergence of the meridians may also mislead the seamen. Magnetic Disturbance caused many times technical and scientific errors that offered feedback for the advancement of science. Among the various studies of electricity and magnetism, some of them related shipwrecks to magnetic disturbance caused by deposits of iron ores. Magnetism, however, is not only an *explanans*, but also a means of discovery, such as with magnetometers.

In the case of storms also, rare weather phenomena, such as the St. Elmo's fire, surprised Magellan and many other sailors. Moreover, not only storms but also still waters might trap a sailing ship for days or weeks without movement, without observation of the longitude in the horse latitudes of the western portion of the North Atlantic Gyre, the great rotating system of ocean currents that represents the interaction between the doldrums, trade winds, horse latitude, and westerlies in the North Atlantic.

Similarly, the danger in the Drake Passage or Mar de Hoces, but also all around the Southern Ocean, could not be completely explained before the comparison of a series of physical phenomena related to the *Antarctic Circumpolar Current* and the southern oscillation. Similar significance retains the north-Atlantic oscillation and the El Niño-Southern Oscillation phenomenon in the Pacific Ocean (Fernández-Armesto, 2004).

3.5. The rise of the Atlantic World

The Early Modern Era was marked by a specific transition from the mythical and vague cognitive behaviour, to the scientific and technological production. Whereas China was considered, until 1500, as the main centre in the domains of engineering, navigation, printing, political organization etc., after 1497, the Maritime Revolution, the proliferation of the spice fleets, and the installation of transoceanic colonies, created a qualitative shift in world history. The oceanic explorers and the founders of the New World let also sharp contradictions develop between absolutism and liberalism, imperialism and national liberation, religion and Enlightment.

The amalgamation of maritime history with world history triggers methodological disputes concerning not only the alleged natural, economical or societal causality behind the rise of global trade and colonialism, but also the subjective cultural content of the enthusiasm for orientation. The invention and the implementation of machinery and tools for exploration was not simply a result of technical accumulation, but an organised enterprise which was enriched and accelerated by the achievements realised in the space-time of the discoveries.

The rational advancement, however, was gradual, painful and puzzled. Christopher Columbus learned the wealth and the location of the Eastern lands from the Florentine Paolo Toscanelli. In his first voyage, Columbus was looking for Japan and he supposed that the shores of Cuba were the mainland of Cathay. Only in his third voyage he suspected that the South American coast could be a large, unknown continent. But in his last, fourth journey, Columbus mistakenly took Central America for Indochina.

However, geographers of his days did not agree with Columbus's views: "In 1494 Peter Martyr introduced the concept of a 'Western Hemisphere', and in 1496 Columbus's friend Bernáldez told him that another 1200 leagues sailing westward would still not have

brought him to Cathay" (Skelton, 1958: p. 59). The possible consequences of mistaken conceptions of the sea-routes could have been
unexpected, as in the case of the false belief on the existence of a
short northwest passage among the islands, of which the New
World was still supposed to consist.

The ingredients of the communication formulas between explorers and natives referred not only to the mutual understanding of
their desires, goals, and actions but also to a universe of beliefs,
nested beliefs, skills, false beliefs, and pretentions. Nevertheless,
apart from the cultural practices, the scientific methods were necessary. The Portuguese managed to observe the latitude by taking
sights of the Southern Cross. The cosmological triangulation culti-
vated a shared attention between the explorers, which created an
inventive behaviour that incorporated also dangerous theories,
such as the existence of the void. The discoveries resulted from the
axiom that knowledge comes from seeing, traveling and experiencing; the gaze of Humboldt (1858) to the civilizations of Cosmos
wielded from the closed to the open universe and created a great
tradition of scientific research approaches.

The age of the oceanic discoveries was a period of significant migration, interaction, quest and competition for wealth. In a fragile
post-colonial world, inherent differences in wealth and production,
between regions, such as Peru and Rio de la Plata, were remarkable.
Brazil became gradually a precious source for information upon
natural history, medicine, geology, mineralogy, and geography.
Alexander von Humboldt and Herbert Ingram Priestley presented
the civilizations of the 'Ancient Inhabitants of America', e.g. the
Toltecs, the Cicimecks, the Acolhuans, the Tlascaltecks, the Aztecs,
and their cultural and historical edifices, for instance, the Mexican
teocallis, houses of gods with pyramidal form. By the encounter of
different civilizations, cultural asymmetries occurred in Latin America, such as with the use of the *khipu* knotted cord systems for the
representation of a decimal system for tribute and recordkeeping in
the Inca Empire (Kenney, 2013).

The Dutch, after 1595, using Linschoten's sailing directions,
travelled to the East and traded helmets, armour, weapons, glass,
velvet and German toys. The foundation of commercial companies,
such as the Vereenigde Oostindische Compagnie (Dutch East India
Company) in 1602, was another consequence of the oceanic discoveries. The antagonism between Spain and Netherlands com-
pelled Dutch merchants to develop trading from the Baltic area and

the Atlantic, to Russia, Italy, West Africa, America and Asia. Their traditional shipments of grain, salt, herring, and wine, were gradually supplemented with luxury textiles, sugar, metals, jewellery, weapons, spices, etc.

This expansion resulted to the dissemination of the part-ownerships or share ownerships, which had a decisive influence in financial sphere (Adams, 1996). In the early seventeenth century the Portuguese had reached Nagasaki and the Dutch were based in Hirado. The VOC was also active in privateering operations against Portuguese, Spanish and neutral ships of the main Manila and Macao routes. The San Antonio incident refers to the first prize captured by VOC in the Shogun's waters (Clulow, 2006) when the Dutch colonialism was replacing the Portuguese in Malacca (1641) and Ceylon (1658).

However, the network structures between principals and agents proved to be vulnerable by uncertainty and infighting problems that undermined the Dutch hegemony and opened the way for the rise of the English empire. The English East India Company, founded in 1600, had headquarters in Bombay, Madras, Calcutta, etc., headed by a Governor General or President and a Council of senior merchants. Yet, the English metropolis maintained the central position in the English East India Company, while VOC was controlled by the Batavian High Government.

In 1611, the Dutch captain Brouwer had introduced a quicker route to the East Indies, through the lower latitudes of the Roaring Forties and then north to the Great South Land. Almost one-fifth of the VOC shipwrecks, that is, around fifty wrecks in Malacca, Gabon, St Helena, Mauritius, Cape Town, Skilly Islands etc., from 1606 to 1795, have been discovered. Australia's earliest known shipwreck was found in 1969 and belongs to the English East India Company ship Trial. The shipwrecks of Trial in 1622, Batavia in 1629, Vergulde Draeck in 1656, Zuytdorp, Cervantes, and Georgette have been identified in Western Australia.

The Northeast Pacific Coast was isolated until the 1770s when cargoes of fur pelts started sailing from Vancouver Island to the entrepôt of Macao. Spain was fragile in the Pacific; the Dutch warred on the Portuguese. Japan was open for Dutch, Portuguese, and English traders, but the Shogunate controlled the marketing of Chinese silks (Gough, 1992).

Conclusions

The installation of trading colonies along the routes of America, Africa, India and Asia presupposed the utilization of the widest range of scientific knowledge, natural resources, economic planning and political thinking. With the great discoveries of Abel Tasman and Willem Schouten, scientists like Bernhardus Varenius were attracted to geography and investigated mathematical data in Earth's motions and dimensions, the solar affection to the earth, the stars, the climates, the seasons, map-construction, longitude, etc. The voyages of James Cook combined the colonization plans with the scientific research, but also with commercial, industrial, transportation and military purposes. Captain James Cook, in his third voyage to the Pacific (1776-80) found in Tonga rigged canoes.

The achievements of the navigators of Oceania, Egypt, pre-Columbian Ecuador, Putun Mayas, are significant instances of long-range maritime trade before the age of global sailing. Furthermore, Native North Americans had developed various types of vessels, such as kayaks, canoes, umiaks, planked boats, and baidarkas. Aside from this, before the eleventh century, Persians and Arabs sailed from Baghdad directly to Guangzhou, according to recent archaeological evidence from a wreck discovered off the coast Belitung, Indonesia (Gaynor, 2013). The Chinese approaches to the Indian Ocean were not rare. Until the fourth decade of the fifteenth century, Chinese junks visited the harbors of the Persian Gulf and the Red Sea, and between 1405 and 1433 large Chinese armadas voyaged from Java to Malindi in East Africa. "As late as 1621 a Chinese collection of charts and sailing directions was published for the voyage from Nanking to East Africa" (Skelton, 1958: p. 16).

Therefore, the transoceanic world is a moving, transforming world with various intercultural influences and interactions. That is to say, any economic-historical analysis on the basis of the demand for sugar, spices, and bullion, should also be extended to the critical contradiction between the world-level exchanges and rivalries, the globalizing treaties, the *Mare Liberum* and the transcultural and dynamic character of the oceanic discoveries.

Chapter 4

Material science and culture

During the first centuries of the Middle Age, not only the feudal local communities but also the majority of the urban settlements were based upon personal consumption. Once global maritime transportation opened the trade of raw materials for the heavy mining and metallurgical industry, production for the purpose of exchange began to emerge. The minting of the gold *genoin* in Genoa in 1252, the gold coin produced in Florence the same year, followed by the new gold coins in Venice in 1282 and Siena in 1333 had raised the market price of gold and opened the way to the substitution of the golden bezant of the Byzantine Empire. The social conflict, as it was reflected by the peasant insurrections, was widely diffused "in northern Italy and then in coastal Flanders at the turn of the fourteenth century; in Denmark in 1340; in Majorca in 1351; the Jacquerie in France in 1358; scattered rebellions in Germany long before the great peasant war of 1525" (Wallerstein, 1974: p. 24).

In the fourteenth century, the economic crisis was severe, while the society was paralysed by a voluminous superstructure, unbearable to the available productive resources, resulting thus in *Wüstungen,* wars, enclosures and depopulation. During the fifteenth century, however, the development of mining and industry in Europe increased the demand for gold, silver, iron, copper and other materials. The growth of exchanges in the next centuries, which was caused by the discoveries, rendered the demand for precious and industrial metals enormous.

The unbalanced trade with the East was draining the silver from Europe, to exchange it with spices, silks, furs, etc. The terrible 'bullion famine' could not be compensated until the age of global sail, when Columbus's successors found rich deposits of silver in Peru and later in Mexico. The exploitation of the Peruvian silver mine of Potosi became easier after the discovery of mercury (1563) at Huancavelica in Peru. After that, the production of silver in the whole of America reached 300 tons a year, while 170 tons a year were sent to Europe.

The Spanish Crown taxed away one-fifth of the silver and the rest was brought to the mint at Seville, while Dutch, French, and English pirates captured some of the loads.

The Spanish had to pay out the Genoans, the Fuggers and other German bankers who supported financially the discoveries. In parallel, a price revolution occurred, which quadrupled the prices from 1411 to 1560, correlated with the bullion famine and the delirious search for money, the discovery of silver, the devaluation of the currency, and perhaps with squandering and coinage of plate. War and silver "leapfrogged, and war is one of the greatest strains on resources and contributors to inflation," according to Kindleberger (1984: p. 29).

An indicator of the transition to the Early Modern History was the transformation of the armoured soldier by the introduction of cannons made of bronze and brass, handguns, muskets, in other words, the coming of the gunpowder revolution, after the fourteenth century (McNeill, 1982). A related benchmark may be the last victory of an English army using the longbow at Flodden (1513).

The use of gunpowder presupposed resources, technical knowledge, and equipment. Apart from the Chinese and the Muslims, the first Europeans that described saltpeter mixtures were Roger Bacon ("Epistola de secretis operibus artis et naturae"), Mark the Greek and Albertus Magnus. The knowledge upon saltpeter led also to the discovery of nitric acid (Biringuccio, 1943; Williams, 2003).

Nevertheless, the great oceanic discoveries interconnected with technological advance, global trade and exchanges. The rise of a world economy was based on geographical expansion and agricultural innovation, on the development of differentiated methods of labor control and on the creation of robust state structures. The beginnings of the industrial revolution may be found in the coal industry, the blast furnaces and forges located in woodlands, where water power, limestone, and iron ores were accessible. The British woodlands were also very important for the construction of the ships (Nef, 1932; Brinley, 1993), but very soon, oak and fir trees, together with pitch and tar, should be imported from Baltic lands.

4.1. Mining engineering and mineralogy in Early Modern Europe

The *Schedula diversarum artium,* compiled around 1100 by a Bene-dictine monk under the pseudonym Theophilus Presbyter, was a manual upon the production and use of painting materials, glass making, goldsmithing, and metalworking. Industrial mills for me-tallurgical applications were documented in southern and eastern German regions, in Harz Mountains, Saxony, and Bohemia, while the eastern influence was widespread. Metallurgy was an important part of the so-called 'medieval industrial revolution', supported by human-labor saving innovations such as the cam, the crank, the trip-hammer, the copper mill, the boring mill and the hydraulic pump. Metallurgical processes were widely adapted to waterpower in Spain, France, England, Germany, Sweden, Poland, and Italy during the thirteenth and fourteenth centuries.

> *The earliest well-established examples of forge mills are at Kirkstall Abbey in England (circa 1200), Évreux and Evry in France (1202 and 1203), and the village of Toaker in Sweden (1224). Other metallurgical ma-chines, such as waterpowered bellows and pumps, ap-pear to have first emerged in Italy in the early thir-teenth and fourteenth centuries respectively. There is reason to believe, however, that forge mills and waterpowered bellows may have been present in Spain and France a century or more earlier, although it re-mains unclear whether the machines concerned were independently reinvented by medieval Europeans or were adaptations of Chinese and/or Roman technology from Islamic Spain and North Africa* (Lucas, 2005: p. 22).

Vanoccio Biringuccio, the author of *Pirotechnia,* the first printed book on metal processing, presented the properties of gold, silver, copper, lead, tin, iron, the production of steel and brass, and the 'semiminerals', such as quicksilver. He reported various methods of drawing wire of gold, silver, iron, copper and brass, and he noticed that wire was drawn of every metal except tin and lead. Biringuccio also discussed how gold was hammered and soldered over copper bars. He also engaged in military activities, namely in the casting of cannons. Biringuccio referred to Agricola's *Bermannus,* he com-pared minerals from Saxony with ancient ones, and described assaying, smelting, casting methods etc. Agricola later reproduced Biringuccio's knowledge:

> Agricola's 'refreshing of his memory' consisted of copying in extenso, without further acknowledgment the earlier author's accounts of mercury and sulphur distillation, glass and steel making, and the recovery by crystalliza-tion of saltpeter, alum, salt, and vitriol together with other less important sections (Smith, 1943: p. xvii).

Despite these critics, Agricola's work retains its own value, as it is shown by the description of the various processes of cupellation, cementation with saltpetre, liquation with the use of lead, amalga-mation with mercury, refining etc. The reader will admire the woodcut illustrations of Blasius Weffring of Joachimsthal, the min-ing town where Agricola lived before he moved to the larger town of Chemnitz (Long, 2003). Nevertheless, the most significant Renais-sance author on precious metals and metallurgy is Lazarus Ercker, whose *Treatise on Ores and Assaying* (Prague, 1574), was translated into English in 1683.

Mining engineering presupposed "different theoretical and prac-tical branches, including chemical mineralogy, natural history, physics, geology (*Lagerstättekunde*), mine surveying, mechanics, hydraulics, and hydrostatics" (Dym, 2005: p. 833). Mining and qua-rrying, iron, glass, and ceramic manufacture were dependent to each other. Water power was introduced in various practices, such as smelting, i.e. melting or fusing ores in order to separate the me-tallic constituents. Forges, blast furnaces, and blooms, minting, i.e. manufacturing of coins, hardening of armour, etching, slack-quenching, finery, were widely disseminated.

In the Early Modern years, "the most advanced German metallur-gy appears in Augsburg and when German armour becomes more readily identifiable through the use of marks" (Williams, 2003: p. 361). Moreover, Westphalia, Innsbruck, Prague, Landshut, and especially Nuremberg produced famous armours. The adjacent Palatinate (Oberpfalz) and later Styria was supplying Nuremberg's armorers' craft with iron and steel, while the Emperors Maximilian I and Charles V, the Estates of Styria, the Council of Solothurn etc. ordered large quantities of armour made in Nuremberg.[5]

> *This metal went first of all to the hammer masters that operated the trip-hammers powered by water-mills which turned blooms or billets into plates ("Blech", or "Zeug") for the armourers. The hammer masters were a craft with their own regulations, that specified that the masters from the hammers at Dut-zendteich, Lauffenholz or other places, were to prom-ise to make the metal for the harnesses of "at least half-steel or better" (zum wenigsten halbstählern oder besser) and nothing less, and to sign it with the city mark or their own mark* (Williams, 2003: p. 590).

The economic renaissance of France, in the second half of the fifteenth century and at the beginning of the sixteenth, was based on the surge of production in the iron forges of Berry, Nivernais, Champagne, Burgundy, Normandy, Brittany, Vendée, Limousin, Périgord, Pyrénées, Forez and Dauphiné. "According to the papers of Chancellor Poyet dating from the early 1540s, there were some 460 iron forges and foundries in the kingdom, of which 400 had been created in the past fifty years" (Heller, 1996: p. 16).

Coal-mining was also developed because coal was essential as fuel for the ironware and firearms industries. In addition, the mas-

---

[5] Until the middle of the twentieth century the study of metallurgical literature was incomplete, although significant information was available from studies of the standard histories of chemistry. J.R. Partington in his *Origins and development of applied chemistry* (1925) collected practical knowledge from ancient and modern literature on the tradition of European metallurgy. Historians of science and technology also offer valuable instruction such as G. Sarton's monumental *Introduction to the history of science*, and L. Thorndike's *History of magic and experimental science*. Furthermore, W.B. Parsons' *Engineers and engineering in the Renaissance* conveyed an excellent picture not only of the state of engineering and architectural science, but also of the general conditions under which sixteenth-century technologists worked.

ters of silver, gold, zinc, copper and other mines, introduced "sophisticated machines like pumps, mine-shaft elevators, wooden tracking, water-driven hammers, bellows, kilns and furnaces" (Heller, 1996: p. 19).

4.2. Studies of magnetism

The coincidence of wonder and order in the magnetic properties was highly influential in the age of the discoveries. A vivid illustration of the wonderful magnetic mystery was given by the impossibility of separating the poles of a magnet. Terrella, Verticity, Electrics, Magnetized Versorium, Non-Magnetized Versorium, Armed Loadstone, Meridionally, Cuspis, Crotch, Cork, Radius, Sphere of Influence, Sphere of Coition, Ostensio, Magnetic Coition and Declinatorium are some of the terms used by William Gilbert of Colchester in his work *De Magnete* (1600). Magnet was also used as a metaphor in the history of medicine. Paracelsus tended to regard magnetism as identical to imagination (Schott, 2002), while Kepler was influenced by the works of his contemporary William Gilbert. The first instances of direct evidence, as mentioned by William Gilbert, were Sebastian Cabot's discovery that the magnetized iron (needle) varies regarding the longitude and Gonzales Oviedo's remark that in the meridian of the Azores there is no magnetic variation. William Gilbert (1893) distinguished also the discovery of the dip of the magnetic needle by the navigator Robert Norman.

The Loadstone, the Magnete, the Sideritis, Ferrarius lapis, Magnes Herculeus, Aimant, Piedramant, Calamita, Siegelstein, were incompletely described and studied before the age of global trade, when all manner of commodities, such as stones, woods, spices, herbs, metals and metallic wares, were avidly sought for, all over the earth, as Gilbert (1893) wrote. Iron and, especially, the best quality of iron, which is called *chalybs*, is attracted more strongly and readily by magnets; while the covering and smearing of iron with greasy fluids delays the magnetic affection. It must also be used for the protection of magnets in iron cases.

Iron ores also, when they are rich, attract iron ores, but not as readily as magnets. Wrought iron attracts wrought iron, as well. Besides, iron is everywhere in the world, mainly in mountains, abundant. The mining of iron was the most important industrial activity for millennia, in many different regions and countries, such as Pontus, Andria, Palestine, Carmania, Britain, Meroe, Cantabria, Gaul, Germany, Crete, Euboea, Pyrenees, etc., as Gilbert (1893) already knew.

4.3. Magnetism and electricity

The ancient Greeks knew that amber displays electrical properties, but Gilbert observed that glass, sealing-wax, sulphur, and precious stones, also attract pieces of paper and straw when rubbed. Gilbert noticed also that the magnetic forces act only on loadstone and iron, orientating them in a particular direction, whereas the electrical forces act upon many different materials and they are non-orientating.

Benjamin Franklin's research was based upon the studies of magnetism by William Gilbert at the end of the sixteenth century and on the discovery of the electric shock and the Leiden jar by Pieter van Musschenbroek in 1745. Johannes Kepler (1571-1630), Athanasius Kircher (1602-1680) and Johann Baptist van Helmont (1579–1644) made important researches upon magnetism.

After the invention of the voltaic pile in 1799, Hans Christian Ørsted in 1819-20 observed the magnetic deflection. Arago discovered the magnetic attributes of a galvanic current passing through a copper wire. In 1820, Ampére detected that currents passing in the same direction attract each other, while repelling in opposite directions. At the same time, Johann Schweigger produced the galvanometer. On September 3 and 4, 1821, the electromagnetic rotation produced by Michael Faraday's electric motor corroborated Ørsted's 1820 discovery of the magnetic effects of an electric current.

A permanent magnet is cemented vertically in the center of a mercury bath. A wire, with one end immersed a little into the mercury, is suspended over the magnet in such a way as to allow for free motion around the magnet. The suspension of the wire is such that contact can be made with it and one pole of a battery. The other pole of the battery is connected to the magnet that carries the current to the mercury bath, and thence to the other end of the wire, completing the circuit.

The apparatus produces a striking phenomenon: when an electric current is run through the wire, via the magnet and the mercury bath, the wire spins around the magnet. The observed behaviour of Faraday's apparatus requires no interpretation. While there was considerable disagreement over the explanation for

this phenomenon, no one contested what the appara-
tus did: it exhibited (still does) rotary motion as a con-
sequence of a suitable combination of electric and mag-
netic elements (Baird, 2004, p. 2).

Faraday observed also the rotating effect of magnets to the plane
of polarization of light and the magnetic influence to the orientation of crystals (Harman, 1994). Therefore, he conjectured the
existence of lines of magnetic force and hypothesised the existence of *ether* as a means of transmission. William Hyde Wollaston
used the laboratory of the Royal Institution for experiments upon
magnetic rotation.

William Sturgeon, in 1825, constructed the electromagnet by
bending the bar, or rather a piece of iron wire, into the form of a
horse-shoe, covering it with varnish to insulate it, and surrounding
it with a helix of wire, the turns of which were at a distance. Joseph
Henry in 1829-30 produced the intensity or spool-wound magnet.
He was also the first who managed to magnetize a piece of iron at a
distance, producing this way telegraphic signal. He developed the
relation of the intensity magnet to the intensity battery, and their
relations to the magnet of quantity, and demonstrated that the
intensity battery and the electromagnet should be employed in
telegraphy (Smithsonian, 1880).

The discovery of electromagnetic induction by Michael Faraday in
1831 raised the problem of how the electromagnetic forces are being transmitted. In 1845, Faraday first introduced the term *magnetic
field*. In 1852 he stated that the magnetic forces can relate to each
other only by means of dynamic curved lines through the surrounding space. Faraday discovered that bodies such as glass, blood,
nitrogen etc., were repelled by the magnets; thus he called those
bodies diamagnetic. In October 1851, Faraday investigated the
electrical current produced by rotating a magnet in relation to a
residual wire connected to its poles, and then the converse, rotating
the wire around the residual poles of the magnet (Tyndall, 1961).
During the 1850s, Kelvin and Maxwell preferred the concept of the
field, instead of the assumption of "forces acting at distance". Humboldt believed that there are intimate interconnections between
various physical phenomena such as the internal heat of the earth,
the molecular forces, and their electromagnetic results. Humboldt
doubted about the origin of the magnetic storms, whether they are
terrestrial magnetism, as Friedrich Gauss insisted, or atmospheric.

Humboldt also emphasized the value of magnetism as a means of measurement and investigated the magnetic power of telluric substances (Von Humboldt, 1858).

Like the *parallel* magnetic alignments, the right angle and elliptical magnetic arrangements are also very common in nature, especially in the case of electromagnetic phenomena. The crucial observations were the difference between electrostatic and electromagnetic field and the alternative manifestations of electromagnetism as either electric or magnetic field. Remarkable was also that the electromagnetic forces are completely depended upon movement of the wire in relation to the magnet.

Kelvin imagined ether as plenum and represented the field of forces with a mathematical constant in the ether. In the late 1850s, Kelvin showed preference to the theory that the ether must be considered as a fluid. In the decades of 1850 and 1860, Maxwell formulated a series of physical and mathematical theories of the field, relying on the physical concepts of Faraday, and on Rankine's idea of molecular vortices, which was eventually shared by Kelvin. This mechanical model represents the function of field during the transmission of electric power, through the action of some particles into the ether. The speed of the electromagnetic waves through the ether coincides with the speed of light.

To explain the intensity of the forces acting on the Physical Lines of Force of the field, Maxwell assumed the existence of an incompressible fluid, moving through the lines, but he stressed that this fluid, of course, is not real but only a natural analogy. These perceptions of ether and field, proposed by Faraday, Kelvin, and Maxwell, have a limited relationship in regard of the terminology with that of *weightless fluids*, which in the eighteenth century were considered as having physical existence. A closer resemblance was the elasticity of the magneto-electric medium. In electrostatics, the elastic deformations of the mechanical ether corresponded to the electrostatic field, coupling thus the universal mechanistic theory of the Maxwellian field. Moreover, the famous experiments of Heinrich Rudolf Hertz were conducted to provide an answer to a problem pointed out by Helmholtz: the problem of experimental control of the relationship between polarization and electromagnetic phenomena. Hertz had two contributions to the development of field theory: the direct verification of the propagation of electromagnetic waves, and a radical critique of the conceptual structure of the field equations of Maxwell's Treatise. Hertz reformulated Maxwell's equations for electromagnetism. Maxwell had focused on optical phenomena. He had proposed the electromagnetic ether as the medium that propagates light, and apparently, he did not imagine the possibility of a direct experimental detection of electromagnetic waves. Frequencies lower than those of the light did not occupy him.

Hertz showed that Maxwell's assumptions about the dissemination of electric power were consistent with the known laws of electrodynamics, but he had not considered a hypothesis regarding the physical existence of the ether. He mainly studied the electromagnetic phenomena and he gradually concluded to the concept of the electromagnetic ether. He measured experimentally the wavelength of the electric wave, and by means of the predetermined frequency of the oscillator (the induction coil), Hertz estimated that the speed of electric waves equals the speed of light.

Hertz focused electric waves with mirrors, to show their reflections. He refracted them through a prism of tough pitch. He produced, finally, polarization phenomena by using metal racks. Hertz concluded that these experiments demonstrated the identity of light, radiant heat and electromagnetic wave motion. Therefore, thanks to the experiments of Hertz, Maxwell's theory was widely accepted. At the end of the nineteenth century, the interaction between magnetism and electricity was being studied with a deep interest while provoking premature questions such as upon the alleged *incompressible* nature of electricity (Hall, 1997).

Magnet coils, magnet cores, magnet frames, magnet yokes, magnet poles, magnet steel, tungsten and cobalt, alloys etc. constituted some of the main parts of the instruments used in the applications of magnetism. Relevant key theoretical concepts were the magnetic field, the magnetic axis, the magnetic bearing, the magnetic blowout or blow-out coil, the magnetic braking, the magnetic circuit, clutch, compass, component, creeping, damping, declination, deflection, detector, difference of potential, flux, moment etc.

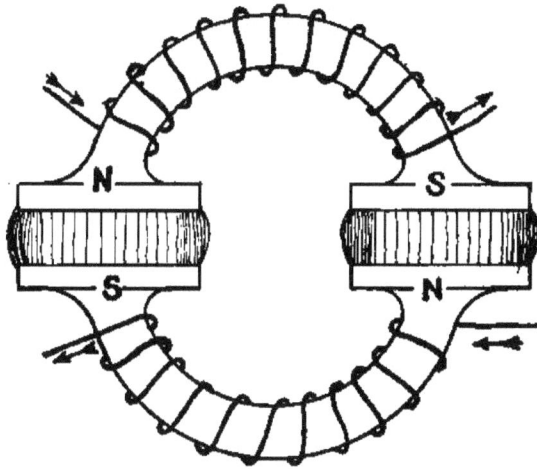

4a Magnetic Pole-Piece. In: Rankin Kennedy, C.E. (1902). Electrical Installations of Electric Light, Power, Traction and Industrial Electrical Machinery. Vol I. London: Caxton.

Moreover, minerals, such as iron, steel and nickel, especially the magnetic oxide of iron (oxide of iron or magnetic iron-ore, Fe_3O_4) and the magnetic pyrite (pyrrhotite), and generators such as the *magnetron* are critically significant for the acquisition of experimental evidence (Tweney and Hughes, 1958).

Modern physical research has made immense progress in magnetic researches. For instance, significant contributions refer to the quantum theory, which studies the strong magnetic forces of certain materials that are called *ferromagnetic.*[6] Theoretical questions, however, remain open, as for the origin of the geomagnetic energy, which is supposed to come from the iron core of the earth, while the mantle consists of silicate minerals which are insulators (Pielou, 2001). Solar activity and sunspot cycles are connected with geomagnetic storms, as well. As a matter of observation of geomagnetism, the annual mean in H is smaller in sunspot maximum years (Vestine et al. 1947).

---

[6] According to this definition, the ferromagnetic materials are consisted of domains of electrons, regarded as miniature magnets (Pielou, 2001).

Chapter 5

The petroleum pioneers
in the age of illumination

The clarification of the historical development of various petro-leum technologies is the topic of this chapter. We will discuss the early oil processing methods, from the refining of coal oil to the introduction of gasoline and lubricating oils, along with the relevant technology transfers and improvements, until the invention of the reforming refinery method, the catalytic or thermal cracking of naphtha etc.

Wood tar distillation is an ancient process. During the distillation in glass stills, it was observed that the distillates differ according to temperature and duration. The dry distillation of wood was employed in three or four receivers, to produce tar, pitch, essential oils, and gaseous products. The heavy fraction of resinous wood was used for sealing ships, textiles, roofs, windows etc. The gas was condensed to oil and tar. The oil was dissolved into light oil and turpentine. Innovations were gradually introduced with the thermal insulation of the stills (Raimond Lullus), steam distillation (Claude Dariot), separation of the heating source (Johann Glauber), air cooling and water cooling. In the years 1792-1807, Philippe Lebon and William Murdoch solved the question of producing gaslight with wood and coal dry distillation. In 1813 Jean-Baptiste Cellier Blu-menthal obtained a patent for the first continuously working distillation column.

Whale oil was the basic source of oil from the sixteenth to the nineteenth century, and it was used experimentally by Benjamin Franklin. At the end of the eighteenth century lard oil was used for the Cape Cod lamp. Between 1830 and 1850, the need for more and better light led to the distillation of illumination oil from turpentine (camphene) and to pioneering work in the production of coal oils, which stimulated the research upon: a) materials for the enrichment of manufactured gas, and b) mineral lubricants by means of low temperature distillation.

A.F. Selligue of France acquired a patent in 1834 for producing water gas, a mixture of carbon monoxide and hydrogen gas produced by reacting steam with incandescent coke (Ayres, 1990). In order to carburet and enrich the water gas, he used oil produced by distillation of Autun schists (shale). In 1847 petroleum seepage was discovered in a Derbyshire coal-mine, yielding 300 gallons per day.

In the early 1850s, the technical chemist James "Paraffin" Young (1811-83), assuming that petroleum had been condensed from coal, undertook the development of the required refining technology of oil-bearing coals and shales. James Young of Scotland produced paraffin oil from coal (paraffin-rich Boghead coal from mines near Bathgate, Scotland).

> *Young obtained a patent in England in 1850, and in the United States in 1852, which he described as "improvements in the treatment of certain bituminous mineral substances and in obtaining products therefrom." His patent explicitly covered the recovery of paraffin-type crude oils from coal by means of slow, low-temperature distillation... Young emphasized the importance in the initial distillation of bringing the temperature in the "common gas retorts" gradually up to "a low red heat," where it should be maintained until the volatile materials were removed* (Williamson and Daum, 1959: p. 38).

The temperature should not rise above the "low red heat" because in that case the products would be converted into permanent gases. During the distillation process, oil of vitriol (sulphuric acid) was used to remove impurities and caustic soda to remove the acid and remaining impurities.

At the same time, in America, the consulting activities about coal and oil investments were widespread disseminated, as Lucier (2008) stresses. The growth of Gas Light industry in the United States, between 1816 and 1850, generated the need to discover a substitute for coal, as the basis for the production of gas. The reasons were the reduction of the costs, the accessibility to supplies and the impurities in coal gas-sulphur compounds, creosote etc. Whale oil, rosin oil and, especially, cannel-type coals were proposed as alternatives. Furthermore, C.B. Mansfield in England developed and patented a process, in 1847, to obtain benzole, for gas enrichment, cheaply from coal tar, but with no success. It was Abraham Pineo Gesner, the petroleum pioneer and "father of petroleum industry," who refined kerosene from coal, asphalt rock (Trinidad asphalt) and petroleum. Gesner experimented with the distillation of Trinidad pitch and Albert bitumen from mines in the Canadian province of New Brunswick. In August 1846, he gave a demonstration, burning his oil in lamps at a public lecture. In 1849, Gesner developed an improved process to produce, through dry, low-temperature distillation, illuminating gas directly from asphaltum. Another product of asphaltum was kerosene, which was registered as a trademark in 1854: When air was passed through or over kerosene, the kerosene itself yielded a rich, kerosene gas. As a whole, the possible range of asphalt products were mineral naphtha, railway grease, hydraulic concrete, mineral pitch, paraffin, coke, ashes for manure, even uncondensed gas for lighting manufacture, and above all kerosene.

In 1852, in Boston, the Atwood brothers made a lubricant from coal tar (a mixture of coal oil, cheap animal fats, and vegetable oils). The advantage of this 'coup oil' was that once blended with fatty oils, it made them more fluid and less liable to gumming through the absorption of oxygen. Atwood experimented also with a few gallons of naphtha from Young's Bathgate plant, which "was mixed with coal-tar naphthas to dissolve rubber for mackintoshes and other waterproof goods. Atwood discovered that by redistilling the paraffin-based naphtha, he obtained a water-white oil which burned in a modified 'moderator'-type lamp with a brilliant flame and without odor or other objectionable features" (Williamson and Daum, 1959: p. 51).

Crude coal oil like crude petroleum required distillation to separate valuable oils from the crude, and distillation combined with chemical treatment in order to eliminate impurities impairing odor, color, burning, and other qualities. The first stage was to charge the crude oil into a cast-iron still and to light the fires underneath it. The vapors from distillation escaped into a pipe, where they were condensed into liquid, and then passed into tanks. The lighter distillates, produced until 600° F, were routed to receiving tanks in order to be further processed into illuminating or burning oils. The rest of the increasingly heavier oils were piped to tanks for paraffin-oil lubricants. The next stages were treatment with sulphuric acid, caustic soda and water washes, and redistillation. In the 1850s, Ignacy Łukasiewicz and Benjamin Silliman Jr. boosted the quality of the processing methods of kerosene.

Gasoline, furthermore, was a by-product of kerosene production. Before the automobile, the refineries tried constantly to get rid the light fraction of crude oil known as gasoline, for safety reasons. They used to burn it for fuel in distilling for oil or let it run into creeks and rivers. Another case of *delay* was the preference for gaslight illumination, instead of the slower disseminated electric light bulb in the 1880s. In 1866, the invention of a more efficient cylinder refining-still improved the process of distillation. By 1875, the refiners implemented a new method for the utilisation of the residuum of the crude oil, left after the manufacture of illuminating oil. In addition, in the shale works of Scotland, they started to refine paraffin wax and lubricants. In 1876, Herman Frasch received a patent for the purification of paraffin wax and waxed paper, and in 1878 for the separation and treatment of oils.

Until the mid-1880s the United States were the world's only producer of crude and refined petroleum. Thereafter the Russian production rapidly expanded, while in the 1890s significant quantities of crude and refined oil came from the Dutch East Indies. In 1882, Nobel brothers, Ludvig and Robert, applied the proposals made by Dmitri Mendeleev to drill oil wells and purify oil with thorough distillation (for example the production of "fotonaftil"). The Nobel brothers established their petroleum company Branobel in Baku. At the same time, Herman Frasch worked upon desulfurizing oil (Herman process), and Vladimir Shukhov developed the method of thermal cracking of crude oil. In the 1890s cheaper kerosene produced from the oil fields at Baku dominated the eastern Mediterranean and the Middle East, due to lower transportation costs.

In 1890 petroleum was discovered in Sumatra, and the Royal Dutch was established. The competition was stronger in the Far East between Standard Oil, M. Samuel and Company, and Royal Dutch. After 1890, Marcus Samuel organized the Tank Syndicate, the forebear of Shell Transport and Trading Company, and supplied the Far East with kerosene from Batum, produced by Rothschild interests.

In the beginning of the twentieth century important was the role of the Anglo-Persian Oil Company and the Burmah Oil Company. After the British oil investments in Peru in the 1890s, the Argentinian oil discoveries in 1907, and the finding of giant oil fields in Mexico in 1910, the large oil companies Standard Oil of New Jersey and Royal Dutch-Shell decided to expand to Latin America.

Early refining technologies

Abraham Gesner's *Practical Treatise on coal, petroleum, and other distilled oils* (1861) offers a historical account of the distillation methods since 1694, with the efforts of Eeele, Hancock and Portlock, and further, in 1761, when oil was produced by distillation of black bituminous shale, according to Lewis's *Materia Medica*, and in 1781, when the Earl of Dundonald obtained oil from coals with dry distillation. Laurent, Reichenbach, and others distilled tar obtained from bituminous schists, while Selligue purified the tars. The main purposes of oil distillation were medicine, lubricating machinery, and burning lamps.

Petroleum distillation was developed by Samuel Kier in Downtown Pittsburgh in the early 1850s. Kier distilled oil in 1854, five years before Drake applied oil drilling technologies to discover significant quantities of crude oil at Titusville, Pennsylvania. The lawyer George Bissell leased 105 acres in Titusville to drill oil, and in 1854 sent a petroleum sample to Prof. Benjamin Silliman Jr., at Yale. Silliman recommended the distillation of the crude oil for the production of kerosene, paraffin, and naphtha. In a few months, Bissell established the Pennsylvania Rock Oil Co., which was later reorganized as Seneca Oil Co. and hired Drake in 1857 to discover oil in Titusville.

Samuel Kier requested the advice of James Curtis Booth, Professor at University of Pennsylvania, Philadelphia. Booth proposed the distillation of crude oil for the production of an illuminant, as a replacement for whale oil that had become expensive. He also gave Kier the drawings of a still. Then Kier, with a wrought iron whiskey still, equipped with a condenser, produced a liquid illuminant, which was called 'carbon oil'. After the introduction of Kier's lamp, the demand for carbon oil increased. Kier built another wrought iron still, with a capacity of five barrels. This second-generation still was a cylinder 42 inches in diameter and 56 inches in height, with a thick circular top and bottom plates with a 12-inch hole cut into the top plate. This still is displayed today at the Drake Well Museum (Mann, 2009). A businessman of New York, A.C. Ferris, experimented with carbon oil, in order to address its deranging odor. He added another refining step, which involved washing with sulfuric acid and caustic soda. Ferris' purified illuminating oil sold 1000 gallons in 1858 and soon became the most popular illuminant in the USA. At the same period, Kier adopted the double distillation, which produced an improved, lighter colored liquid. However,

Abraham Gesner in Canada and James Young in Scotland had already manufactured coal oil, with competing price and quantity. Gesner's patents were sold to the North American Kerosene Gas Light Company, under the denomination 'Kerosene Oil.' Young secured patents in England (7 Oct 1850) and in the United States (23 March 1852) for obtaining paraffin oil and paraffin from bituminous coals (boghead coal).

Charles Blachford Mansfield in 1847 had obtained benzole from the coal tar of gas works. His patent referred to the "purification of spirituous substances and oils" derived from coal tar, etc. He described the following products of coal tar: alliole, benzole, toluole, camphole, mortuole, and nitro-benzole (Gesner, 1861: 9; 93-94). Regarding the development of the refining technologies it was significant that

> *prior to the use of petroleum as an illuminant, successful efforts had been made to distil oil from shales and coal. As early as 1848 patents for such processes had been taken out both in this country and England. The product was called "Coal Oil" and "Kerosene." At the time when Drake drilled his famous well in 1859, there were some fifty such refineries scattered throughout the eastern part of the United States. Six of them were credited to the Pittsburgh district, the largest of all perhaps being that of the Lucesco Company at Kiskiminetas on the Allegheny River. Its distilling capacity was said to be 6,000 gallons per day* (Davison, 1928: p. 93).

> *It was here in Pittsburgh that petroleum was first used as a lubricant and as an illuminant. Here was built the first refinery, the location of which can be viewed from the building in which we are gathered. From Pittsburgh in 1860 went the man, Charles Lockhart, who first interested foreign trade in American oil. Doctor Tweddle, a noted chemist of Pittsburgh in the early 70's made the first lubricating oils by the method of steam and vacuum distillation. His process has not been materially improved upon to this day. From Pittsburgh went the tools and equipment for early Oil creek operations, and today from our mills and shops*

goes the larger part of the equipment necessary for production and refining operations both in North and South America (Davison, 1928: p. 103).

In 1861, Charles Lockhart established the first commercial-scale oil refinery in Pittsburgh, for the refinement of the crude oil produced in Titusville. It was the World's First Major Oil Refinery, established at Brilliant Station on the south bank of the Allegheny River. Cooling water, required for condensing the products obtained by distillation, was conveyed by gravity via pipeline from Lake Carnegie. The refinery capacity was 250 barrels/day. The technology used at Brilliant Oil Works involved distillation, probably followed by treatment with caustic soda, sulfuric acid and finally water washing. Lockhart along with John Gracie, improved refining technology and on 17 November 1863 obtained the U.S. Patent No. 40,632 for improvement of still for petroleum (Mann, 2009). From another viewpoint, the world's first oil refinery was a small plant, set up in the years 1854-56 by Ignacy Lukasiewicz near Jaslo, Poland (then Austrian Empire). The first large oil refinery opened in 1856 at Ploiesti, Romania.

Distillation resources and advances

The most marvelous element in the distillation is the production of brand new substances, such as paraffin and naphthaline that they do not exist in ready form in coal. Abraham Gesner (1861) observed the similarities between distillates of wood and coals, and of coke and charcoal, relating them with the alleged vegetable origin of oil. He regarded the formation of lignite and brown coal as a transition from vegetable to fossil state, by their great loss of oxygen and their increase of hydrogen and carbon. An exception, the so-called south boghead coal, abounds in the remains of fishes and crustacea. Gesner mentioned various materials appropriate for oil distillation, such as coals, bituminous shales, asphaltum, bitumen, bituminous sands and clays, petroleum, lignite, peat, caoutchouc, gutta percha, and the tars produced in the manufacture of stearin. The first effect of the dry distillation of substances that contain oxygen, hydrogen and carbon, is the removal of oxygen from the substance in the form of carbonic acid or water.

After the removal of oxygen, as Gesner (1861) described, hydrogen and carbon escape as carbureted hydrogen, or olefiant gas. The distillation of acids results in the release of oxygen in the form of carbonic acid and water, while new acids are produced. In general, the distillation of coals in a suitable retort produces water in the form of vapor or steam, sometimes together with carbonic acid, ammonia if the coal contained nitrogen, and extremely volatile, light and inflammable hydrocarbons. Later, as the heat increases, oils of different specific gravities are condensed.

When the heat raises to 750-800° Fah., gas, free carbon and various pyrogenous substances appear, which are named *dead oil* that mixes with aqueous products at the bottom of the vat. Usually, all oils in a retort distill over at a temperature of 750° Fah. Coals, coal shales, asphaltums, petroleums and other bituminous substances yield series of homologous compounds, lighter as benzole, heavier as paraffin oil, and solid as paraffin, naphthaline, etc.

Abraham Gesner (1861) made special reference to boghead coal, which was used by Young for the production of paraffin. Boghead coal was also imported in the United States for the manufacture of kerosene at New York and Boston. However, it was quickly replaced by cannel coal and other cheap resources available in the United States. One ton of boghead coal yielded 13,000 cubic feet of gas or in common retorts "120 gallons of crude oil, of which 65 gallons may be made into lamp oil, 7 gallons of paraffin oil, and 12 lbs. of pure paraffin" (Gesner, 1861: p. 21).

Near Petitcodiac River, Albert County, Canada, Albert Coal was mined, similar to asphaltum. It was composed by carbon 85.4-86.3%, hydrogen 8.9-9.2%, nitrogen 2.9-3%, oxygen 1.9-2.2%, ash 0.1%, and traces of sulphur. The average yield of crude oil was 110 gallons per ton. Of the crude oil, 70% could be made into lamp oil and 10% was heavy oil and paraffin. Albert Coal was the richest resource for the manufacture of oils. Lamp oil of good quality was produced by the purification of the Breckenridge cannel coal, which was produced in Kentucky. At a red heat this coal yielded 61.3% volatile matters, 30% fixed carbon, 8% ash, 0.6% hygroscopic mixture, and a trace of sulphur. It yielded 130 gallons of crude oil, of which 58% was manufactured into lamp oil, and 12 gallons into paraffin and paraffin oil. Paraffin was produced in large quantities (143 lbs. per ton) from brown coal, as well. The tar and pitch from the manufacture of stearin were also employed for the production of the so-called grease and candle oil. One ton of candle tar yielded 200 gallons of crude oil, of which 70% could be used for the production of lamp oil and 10% for lubricating oil. Abraham Gesner (1861) referred also to the Bitumen of Trinidad and Cuba, which both produced oils with an objectionable odor.

Kerosene was firstly produced by the New York Kerosene Gas Light Company. The next step was the production of coal oil. According to Gesner, the oily products of the distillation of coal in a retort or closed vessel, at a heat of 1200°, will be small in quantity, including benzole, naphtha, naphthaline, carbolic acid, piccamar, pittical, copnomor, and other hydrocarbons.

However, if the heat does not exceed the 750-800° Fah., "a different class of results follows. Instead of true benzole, eupion will be formed, the naphthaline will be replaced by paraffine, the carbolic acid, copnomor, piccamar, etc. will be less in quantity, and there will be a great increase of the oils employed in lamps and for oiling machinery" (Gesner, 1861: p. 37).

The distillates produced in a heat of 700-800° Fah. were generally oils, whereas those produced in a heat of 1200° Fah. were called tars. The production of illuminating gas requires a temperature of 1000-1200° Fah. The best retorts permit the equal distribution of the heat throughout the coal. For this reason, the revolving retorts or the large horizontal D-shaped retorts heated over a furnace, offered an effective solution. The revolving retorts however were not appropriate for the distillation of Albert Coal and softer bitumens, because they smelt quickly and adhere to the iron. Vertical retorts were introduced in Ireland for the decomposition of peat and were found also in France and other places in Europe. Nevertheless, the cheapest, most economic retort was the horizontal D-shaped, followed by the revolving retort. Very important were also the condensers and, in general, the cooling apparatus.

5.1. Research on petroleum

Early modern science distinguished between asphalt, bitumen, naphtha, pitch black, pissasphaltum, maltha, and petroleum. Pierre Joseph Macquer (1718-1784), George Balthazar Sage (1740-1824), and Richard Kirwan (1733-1812) contributed to the definition of the term *oil*. Kirwan added the value of the specific gravity to the physical properties of the aforementioned substances.

In the beginning of the nineteenth century, petroleum was produced in Hannover, Alsace, Auvergne, Guayana, Austrian Galicia, along with a belt from Krakow to Lvov, in Wallachia, Moldova, Italy, and Russia. In 1802, Genoa, Italy, the municipal administration planned to substitute naphtha for olive oil, to fuel public illumination lanterns. "Lighting oils, solvents and disinfectants, lubricants and waterproofing resins were the main products obtained by a simple boiling of crude oil. Also, petroleum was not considered a necessary ingredient for the production of these goods, but only an alternative raw material" (Gerali, 2011: p. 91). Traditionally, petroleum was a by-product of drilling for salt wells, as the salt was an absolutely necessary preservative.

In 1825, Michael Faraday isolated benzene from the oily residue of the production of illuminating gas and named it *bicarburet of hydrogen*. In 1831, Eilhard Mitscherlich produced it by distilling benzoic acid and hydrate of lime. It may be produced from coal tar naphtha with fractional distillation. It is purified after washing it with sulphuric acid, then with a solution of caustic potash, or soda, and finally with distillation over lime.

The first refiner of petroleum, or rock oil, was Samuel M. Kier of Pittsburgh with his company *Seneca Oil*. He was selling the oil of the salt wells, when a Philadelphia chemist, J.C. Booth, advised him to refine it, to a new petroleum product, lamp oil. Another pioneer refiner was A.C. Ferris of New York, who observed Kier's lamps, distilled lamp oil and eliminated the odor. The so-called "coup oil" of the Napoleon III times, was developed by Luther Atwood and patented in 1853. Notwithstanding its strange odor, it was used by cotton-mill owners and railroad companies. The coal oil production improved the refining methods, because of the demand for satisfactory illuminating oil for manufactured lamps.

In New York, Pennsylvania, and Ohio, the manufacture of lamp oil from cannel coal (coal oil) was diffused in the 1850s. Benjamin Silliman Jr. refined petroleum by fractional distillation in 1854. The next year, 1855, he was hired to write a report for the possibility of using petroleum for illumination. Silliman distilled Pennsylvania rock oil with fractionation and showed that it was a powerful illuminator. His report was a major factor for the development of oil industry, estimating that 50 percent of the crude oil could be used for illumination purposes while pointing out the possibility of paraffin production with high-temperature distillation.

Moreover, in August 1859, E.L. Drake drilled petroleum in substantial quantities in Titusville, Pennsylvania. During 1860 there was an expanding output, between 200,000 barrels and 500,000 barrels, most of which in the latter part of the year. Petroleum production in 1861 was 2,113,609 barrels, reaching three million in 1862, over five million in 1870 and 1871, while in 1873 it increased to 10 million barrels. The same refining methods were employed for coal oil, but the coal-oil industries were gradually abandoned or converted to petroleum refineries. The early refining process consisted in heating crude oil in retorts until it evaporated. "The vapour passed through a worm in which it was condensed, and from this the product ran into distillate tanks. The first product was naphtha, the next was illuminating oil, and the last was a heavy oil containing paraffin. The naphtha was discarded" (Oliver, 1956: p. 342). Illuminating oil was graded by colour; the superior quality was astral white. The process was concluded with sulphuric acid, caustic soda, and other chemicals. Refining improvements had to eliminate foul odors due to the quantities of sulphur in the oil.

In the beginnings of the oil industry, the refineries used horizontal, elevated vats to heat the crude oil and vaporize its volatile components. The vaporization and subsequent condensation of the distillation products was repeated at various temperatures to obtain the separate fractions. The leap from the simple refinery to the introduction of fractionating columns permitted the continuous distillation of the distillates at their different boiling points. The diversification of the distillates increased gradually. Thus, in 1862 the refineries used the process of atmospheric distillation to produce kerosene and naphtha, tar, etc. In 1870, vacuum distillation was introduced, for the production of lubricants, asphalt, residual coker feedstocks etc.

By 1870, the most significant technological innovation was the construction of bigger stills, reaching capacities of 3,000 barrels. A further improvement referred to the bottom of the stills. In 1861, "Joshua Merrill substituted for the conventional riveted bottom of cast or wrought iron, a formed, seamless bottom cut from a single plate of flange iron or steel. The bottom was given a certain amount of curvature in order to keep contraction and expansion from straining the joints where it connected with the sides of the still" (Williamson and Daum, 1959: p. 253). Additionally, Merrill enclosed the entire still in an iron casing, which functioned as an insulating air chamber for better temperature control. The same year (1861) Merrill introduced a pan to hold solid alkali in contact with oil, but in distance from the still bottom. In 1857, Merrill had also introduced the fire-and-oil proof cement for still-repair work. In 1868, Charles Lockhart and John Gracie improved Merrill's air chamber. Moreover, Lockhart and Gracie added in 1863 mechanical scrapers that operated on a rotating shaft which removed the residual from the bottom of the still.

Another improvement was the introduction of steam cooling after four hours from the completion of a run. The cooling steam reduced the time period before the workers could remove the manhole and enter the still. "Steam also expedited the removal of tarry residuum and made it safer to crack deeper to coke. By creating an outward current through the condensing apparatus, it facilitated the removal of dangerous gases and it was also helpful for removing paraffin and other deposits - formerly a leading source of refinery fires" (Williamson and Daum, 1959: p. 255). The increased size of the stills required bigger and longer condensers, pipes and tubing. Gradually, there was a reduction of the sizes, such as with the replacement of the gooseneck exit pipe by a condensing drum at the top of the still.

The development of organic chemistry was a vital presupposition for the invention of refining methods. In 1855, Professor Benjamin Silliman had distinguished between distillation and decomposition (cracking) of various petroleum components. In 1856, W.H. Perkin, Sr., having distilled benzene from coal tar, synthesised from it the dye aniline purple. In 1858, Frederick August Kekulé von Stradonitz distinguished the structural difference of the aromatic compounds, such as benzene, whose molecules are arranged in rings rather than chains. In 1867, M.P.E. Berthelot stated that all hydrocarbons may be cracked or decomposed, on the action of heat, into hydrogen and carbon. The refining process, however, was developed empirically, under certain tasks, such as with the production of kerosene. The use of the term specific gravity was based on the comparison with the unit of the specific gravity of water. By 1862, the refiners measured the densities and the boiling points, although the products with the lowest ones (gasoline, pentane, and butane) had not yet been produced commercially.

The refining technology advanced in north-western Pennsylvania, Erie, Cleveland, Pittsburgh and in nearby cities, after the opening of the Drake Well. Petroleum was discovered in Colorado by Gabriel Bowen in September 1860. The crude oil of Four Mile Creek, northeast of Cañon City, was refined by Alexander Morrison Cassiday in simple stills and sold as lamp oil and lubricant. Until 1900 the crude oil produced in Cañon City and Florence, Colorado, was refined to kerosene that was used as lamp oil. Additional crude oil fields were sought after 1900, to cover the increasing demand for gasoline. Florence, until 1930, was the center of the refining industry in Colorado, producing lubricants and oil for the local consumers.

Seepages near Cañon City were a source of oil in the 1860s, and production from deep wells began at nearby Florence in the 1880s. Local refineries produced lamp oil and lubricants, which the Continental Oil Company, an instrument of the Standard Oil monopoly, distributed throughout the Rocky Mountain region. Natural gas was usually a by-product of oil production until the 1920s, when commercial quantities were produced first in Larimer County, then in western Colorado. During that decade-long time, consumers of coal began using natural gas for space and water heating, as well as cooking (Scamehorn, 2002: p. x).

By the 1890s the coal had supplanted wood in the United States, covering ninety percent of energy demand, while the rest was consumption of natural gas and petroleum. Coal was used for heat in house and business, for the production of steam in industry and transportation, and for the production of coke that was an ideal fuel for the reduction of iron ores and precious metals. The distillation of coal in retorts yielded manufactured gas.

For almost two decades, Florence's refineries were the exclusive source of petroleum products for the states and territories of the intermountain West. Kerosene was extracted by heating crude oil in stills. Lighter fractions, including gasoline, for which there was no

*market at the time, were wasted in order to produce
oil for lamps. About one-third of the crude oil was
consumed in making lamp oil and lubricants. The residual oil and tar were sold as boiler fuel, or burned
under the refinery stills. In the 1890s local and re-
gional ore-reduction mills purchased the residual oil
for roasting ores to remove sulfur, and used tar as boiler fuel. The fuel oil sold for 50 cents a barrel, or
slightly more than 1 cent a gallon. While oil, in this
instance, competed with locally produced coal, the
supply at that time was too small to have a significant
impact on the fuel market* (Scamehorn, 2002: p. 50).

The introduction of the automobile made gasoline useful, while
petroleum industry turned into greater technological improvements in oil refining, such as with the manufacture of lubricating
oil. The leather manufacturer Hiram B. Everest invented a method
for utilising the waste oil from a small refinery to process his
leather, and discovered a new lubricant. The lubricating oils extended "from the light oils to the solid greases" (Oliver, 1956: p.
342). The lubricating methods involved filtration without the application of heat; refrigeration and steam distillation; treatment with
sulphuric acid and lye. Very soon the majority of the nation's railroads were lubricated by petroleum products. After the development of John Rockefeller's enterprise and further discoveries in
Persia, Azerbaijan, and Indonesia, persistently important became
the need for fuel at the projects of the Wright Flyer in 1903, the
Model T in 1908, and the centrifugal gas turbine for jet propulsion
in 1930.

In 1913, thermal cracking was initiated for the mass production of
gasoline, and of by-products such as residual and bunker fuel.
Cracking could increase the production of kerosene by 15-20%. In
1916 the sweetening process was implemented for the reduction of
sulfur and odor and in 1930 the thermal reforming for the improvement of octane number and the production of residuals. The
next innovations were hydrogenation to remove sulfur (1932),
coking to produce gasoline base stocks (1932), solvent extraction to
improve lubricant viscosity index (1933), solvent dewaxing to improve pure point (1935), catalytic polymerization to improve gasoline yield and octane number (1935), catalytic cracking for higher
octane gasoline (1937) etc. (Gary et al. 2007).

The significance of cracking for the production of gasoline

After the invention of internal combustion engine, the vast and persistent increase in demand for gasoline was met firstly by the discovery of additional crude oil, and secondly by the invention of the cracking processes. Distillation and thermal cracking are generally considered as conventional ways of producing gasoline. For many years, refining technique did not permit greater yields of motor fuel. The most important oil refining processes are cracking processes, which convert long hydrocarbons, usually alkanes, into smaller, more valuable chains (lighter alkanes, olefins, branched alkanes, etc.). In 1891, the first cracking process was patented in Russia by Vladimir Shukhov. It was a *thermal cracking* process, namely cracking of the oil feedstock with the use of elevated temperature and pressure, but without the use of a catalyst. The French engineer and inventor Eugene Houdry introduced the *catalytic cracking* process, focused on improved fuels, produced high-octane fuel, and created also an early catalytic converter for cars. In catalytic cracking a catalyst was used, along with heat, to perform the cracking reactions effectively. As a consequence,

> the relative importance of coal had dropped until in 1937 it accounted for 53.7 percent of the total. Much of this change has represented a shift from coal to oil and gas, especially since 1926. In that year the grand total energy supply in the United States was 24,445 trillion British thermal units. In 1937 the total was about the same - 24,183 trillion. During this 13-year period, however, the relative importance of oil and gas rose from 26.2 percent to 42.2 percent (McLaughlin, 1937: p. 123).

The first commercial catalytic cracking unit was built in 1937 in the United States, while in 1942 Fluid Catalytic Cracking (FCC) was introduced, the most widely used process. The FCC process is acid-catalysed. In the 1960's zeolites replaced the amorphous silica-aluminas in catalytic cracking.

The origin of petroleum

The disagreement between the biological and the abiogenic theory of the creation of oil shows that practical use is sparsely related to theoretical congruence. The debate is getting deeper around the

dilemma: a) whether the fuel feeds the fossils, that is, the microor-
ganisms metabolise hydrocarbons, according to the abiogenic theory, b) or the fossils feed the fuel, as the mainstream, biological,
theory contends. The ambivalence is extended, respectively, upon
whether oil seeps up, outgasses to the surface, or migrates down
into the wells. The lack of sufficient evidence is significant to the
continuation of the disagreement. Nevertheless, the core of the abiogenic argument is that the hydrocarbons are not created biologically:

> *Gold claims that when the earth formed, 4.5 billion*
> *years ago, hydrocarbons accumulated as solids. They*
> *have since slowly seeped up through the mantle in*
> *vast quantities to the surface via 'outgassing'. The*
> *presence of hydrocarbons on other planets in the solar*
> *system is claimed by Gold as one of his most telling*
> *pieces of evidence* (Collins and Pinch, 1994: p. 82).

Thomas Gold did not only insist that hydrocarbons are widespread on other planets of our solar system, but he also suggested
that subsurface life may be widespread at depth in the crust of the
earth, as in hydrothermal vents of the oceans and in the universe,
because chemical fluids of the deeper levels of the planets migrate
upward and feed bacteria (Gold, 1992). Furthermore, he pointed
out that the original source of chemical energy for earthly life was
derived not from photosynthesis but from the oxidation of hydrocarbons that were already present, just as they are also present on
many other planetary bodies and in the original materials that
formed the solar system. This chemical energy, according to Gold's
deep-earth gas theory, is created by a long-term and continuous me-
tered supply of chemicals and energy, in regions with liquid water.

5.2. Oil in Arabia

On regard of the oil industry, the role of international relations is
significant. In 1948, Aramco, the largest single investment of private
American capital abroad, discovered Ghawar, the world's largest oil
field, with estimated remaining reserves of 70 billion barrels. The
next year, Getty Oil Company acquired a 60-year concession in
Saudi Arabia's part of the Neutral Zone for 9.5 million dollars. In
1950, Saudi Arabia adopted a fifty-fifty split in profits with foreign
oil companies. At the same time, Aramco completed the Trans-
Arabian Tapline, which carried crude from the oil fields of Saudi

Arabia's Eastern Province to the Lebanese port of Sidon in the Mediterranean Sea. Simultaneously, the global oil production exceeded 10 million barrels per day. In 1951, Aramco discovered Safaniya, the world's largest offshore oil field and third largest oil field, with estimated reserves of 35 billion barrels. In the same year, Iran nationalised its oil industry, dissolving the Anglo-Iranian Oil Company and forming the National Iranian Oil Company.

On July 26, 1956, Egyptian President Nasser nationalised and took over the assets and the responsibilities of the Universal Suez Canal Company. The intervention of the Soviets in Hungary, in August 1956, and the retreat of the French and the Britons in the Middle East, spread fears that the communist influence could also expand to the Arab world. Under those circumstances, the US government issued the Eisenhower Doctrine for the protection of its interests in the area. From here on, the United States regarded nationalism "as an inevitable development which should be channelled not opposed" (Takeyh, 2000: p. 6).

In April 1959, representatives from Syria, Venezuela, Saudi Arabia, Iran, Kuwait, and Iraq signed the secret and non-binding Mehdi Pact at the First Arab Petroleum Congress in Cairo, which aimed at establishing national oil companies, receiving a greater portion of oil revenues, and taking a greater role in determining oil prices. Concurrently, *Universe Apollo* became the first oil tanker to surpass 100,000 dwt. The supertankers appeared in 1953 with the Tina Onassis, which carried 46,080 dwt. The largest tanker reached the 326,000 dwt by the late 1960s. In the mid-1970s tankers of more than 500,000 dwt were built.

Oil from 1950 to 1970 became the main source of energy worldwide and the role of coal was diminished, while half of the total ship-cargos were crude oil and its derivatives. The investments in the oil industry was a diversification factor, which let nontraditional business groups maintain their longevity, by their privileged access to fuel resources. On August 8, 1960, Standard Oil of New Jersey (later Exxon) announced its decision to cut average posted prices of oil unilaterally and without warning by 7 percent, from $1.92 to $1.81 per barrel. Standard Oil of New Jersey was followed by other companies. Oil-producing countries were sent into a rage. Iraq, Iran, Kuwait, Saudi Arabia, and Venezuela proceeded to the formation of OPEC (Organization of the Petroleum Exporting Countries). In 1961, Iraq revoked 99.5 percent of Iraq Petroleum Company's concession area, allowing yet to keep operating its producing interests. Furthermore, a partnership composed of Hunt Oil (headed by Texas billionaire and oilman Nelson Bunker Hunt) and British Petroleum discovered the Sarir oil fields, the largest in Libya.

Oil shocks

In 1963, the Algerian government established Sonatrach as the national oil and gas company. In 1971, Algeria announced, in line with OPEC, the nationalisation of foreign hydrocarbon assets on its territory. In 1973, Saudi Arabia acquired 25 percent of Aramco, while Abu Dhabi acquired 25 percent of Abu Dhabi Petroleum Company and Abu Dhabi Marine Areas under the participation framework of OPEC. Similarly, NNOC acquired 35 percent interest in foreign petroleum operations in Nigeria, whereas Libya nationalised 51 percent of the interests of ENI, Oasis, Exxon, Mobil, and Occidental. Libya also completely nationalised the interests of Amoseas/Texaco, Royal Dutch Shell, Arco, and the partnership of BP and Bunker Hunt.

In the wake of the Yom Kippur War, Persian Gulf OPEC nations raised the price of Arab Light from $2.90 to $5.11 a barrel. The members of the Organization of Arab Petroleum Exporting Countries (OAPEC) began a 5 percent per month cut in collective oil production until Israel agreed to withdraw to pre-1967 borders, and imposed a selective embargo on countries supporting Israel (including the United States). Those actions brought about the so-called First Oil Shock.

The petrol industry suffered now from shortages in supplies and from the higher level of fuel prices due to the oil crisis. The closing of the Suez Canal between 1967 and 1975 was an important benchmark in the global economic history, which affected the dynamics between capesize and suezmax transport vessels. The closing of the canal coincided with the highest freight rates since the end of the Second World War, and the war between Israel and the Arab States, which started on 5 June 1967. The length of the seaborne transport routes of petroleum was tripled, reaching the 12,000 miles. More ships of bigger capacity were now needed because they had to carry crude oil, which was now refined in the place of consumption. After the late 1970s, the bulk cargos of coal were linked more with the production of electricity, than with the steel industry. On the other side, the oil companies, especially after the oil crises, preferred the 'time charters' of tankers instead of ship-owning.

In 1974, most of the Arab oil-producing nations ended their embargo against the United States. Sixteen countries of the Organization for Economic Cooperation and Development (OECD) concluded the International Energy Programme Agreement and established the International Energy Agency, to guard against future supply disruptions by stockpiling and sharing oil. Saudi Arabia increased its stake in Aramco from 25 percent to 60 percent. Kuwait acquired a 60 percent stake in Kuwait Oil Company. Simultaneously, Abu Dhabi, one of the world's largest oil producers (the giant Upper Zakim offshore field in Abu Dhabi has reserves of 98 billion barrels), raised its share in Abu Dhabi Petroleum Company and Abu Dhabi Marine Areas to 60 percent. The companies were then restructured as Abu Dhabi Company for Onshore Oil Operation (ADCO) and Abu Dhabi Marine Operating Company (ADMA-OPCO).

In 1978–1979, the Second Oil Shock was caused by the Iranian Revolution, which increased oil prices to double. Ayatollah Ruhollah Khomeini formally assumed power in Iran in December 1979. Immediately, Iran cancelled the consortium agreement (Iranian Oil Participants). The same year, Egypt was being suspended from OAPEC for signing Camp David Accords. Nigeria nationalised BP's interest in Shell–BP. Shell–BP became the Shell Petroleum Development Company of Nigeria (SPDC), the dominant company in Nigeria's sector (Vassiliou, 2009).

5.3. The international petroleum sector at the end
of the twentieth century

In 1980, the U.S. Congress established the Arctic National Wildlife Refuge. In the biggest part of the refuge, petroleum exploration and development was prohibited, with the exception of the "1002 Area" near the Beaufort Sea. At the same time, Saudi Arabia completed the acquisition of Aramco, while Iran–Iraq War began, after Iraq invaded Iran. Concurrently, Iran acquired the assets of all four successful joint ventures that were outside the framework of the 1954 consortium agreement. The assets were taken over by the new Iranian Oil Offshore Company. In 1981, Aramco completed the construction of the Petroline pipeline from Abqaiq to the Red Sea port of Yanbu. Two years later, in 1983, Ali al-Naimi became the first Saudi president of Aramco (Vassiliou, 2009).

Since 1986, Algeria encouraged foreign participation in oil and gas exploration, considering necessary to attract finance, skilled manpower and the appropriate technology to exploit resources. A 1991 law offered fiscal incentives for the utilisation of enhanced oil recovery (EOR) techniques, allowed foreign companies to work in fields previously subject to the Sonatrach monopoly and lifted the state monopoly on oil and gas transport.

Globally, the environmentalist aspects become all the more important: As a matter of fact, the eightfold increase in the consumption of petroleum since World War II causes global warming, threatening to flood coastal regions and creating havoc with wildlife. At the same time, the nuclear accident of Tokaimura on 30 September 1999, for which the JCO Company was responsible (linked to Sumitomo Metal Mining), urged the Japanese to strengthen the natural gas sector ("Japan: The nuclear accident," 1999).

A period of rising expansions of gas and renewables has already begun. In 1997, Elf Aquitaine planned a Gas Oil refinery opening in Eastern Germany. The same year, Germany imported 43,600 barrels/day Gas Oil from Russia, while in 1996 imported 44.750 barrels/day. Germany imported from the Netherlands in 1997 a quantity of 224,000 barrels/day, from a total of 410,000 barrels/day which was produced by the Dutch. In parallel, a mild increase in demand for diesel for vehicles occurred in Germany, from 453,000 barrels/day in 1994, to 463,000 in 1995. An important difference between France and the EU average is diesel usage rate contrasted to gasoline, which amounted in 1997 to 59% in France compared with 49% in the EU. Overall, the decrease of imports from Russia was due to the economic recovery. The same year, the French company Total was developing a gas field in Sirri, Iran, for sale to Dubai. The agreement provided for a price $ 0.85/mm Btu instead of $ 1/mm Btu pursued by the company. They expected an agreement for 140 million cubic feet of gas per day ("Dubai will precede," 1997).

Saudi Aramco in the first quarter of 1997 was delivering 50,000 barrels/day, reaching thus the second place among suppliers of crude oil, after the producer of sweet crude oil Nigeria. The company aimed also to buy the 27.5% of the state-owned Petrogal refineries to meet all the needs in acidic crude oil. "But Iran, Iraq and Egypt," as the press meant, "supply Petrogal with almost twice the volume of acidic crude oil compared with that of Aramco" ("Aramco withdraws," 1997: p. 3). In 1997 Mobil and Shell prepared in Nigeria a study of the management of Warri and Port Harcourt refineries. The two refineries of Port Harcourt, with a total capacity of 210,000 barrels/day, worked in 1997 by 70%. The refinery capacity of 125,000 barrels/day in Warri was activated only by 70%. Total was then planning to build the refinery in Kaduna ("Nigeria: Delays in implementation," 1997).

In Aghajari region in Iran, there is a huge terrestrial oil field. The first eight months of 1997 occurred a slight reduction of crude oil flow in the wells by 150,000 barrels/day. At the same time, there was an increase in the capacity of domestic refineries by 100,000 barrels/day (rising at 1.3 million barrels/day), of which 80% were using Iranian light crude. Thanks to these two factors, Iran's exports decreased to 900,000 barrels/day (therefore there was a fall of 200,000 barrels/day). In return the NIOC, the state-owned Iranian oil company offered heavy crude oil. The drilling was hindered by the presence of water in the oil and by the chronic shortage of gas for reinjection to the ageing fields, in order to maintain the desired pressure ("Iran: Production Problems," 1997). However, the gas injection projects were postponed because the aquifer was positioned beneath the oil field. These problems had emerged at an earlier stage. In 1997 they began to operate the export refineries in Bandar Abbas, which mainly used other types of crude oil and refined 70,000 tons per day.

In Abadan, in August 1997 there was an increase from 400,000 barrels/day to 430,000. The size of the Aghajari was estimated at 31 billion barrels. It would need about 15 trillion cubic feet for injection to the rest of its lifecycle. The Iranians were saying that the oil deposit was "injured" by the consortium that, led by BP, had begun its development. Similar to the even more complex Gachsaran deposits - which produced about 600,000 barrels/day and whose reserves amount to more than 50 billion barrels - the Aghajari produces only a small amount relative to the size of its content... ("Iran: Production Problems," 1997: p. 3).

The great discovery of deposits in the North Caspian encouraged the Italian ENI and the French Total Fina to support the construction of the Iranian pipeline, which would achieve better prices in the Asian market ("Iran: Increase preferences," 1999) than in the Mediterranean. More than 75% of the offshore Kazakstan International Operating Co. was owned by non-US companies. Transit duties of pipelines reached in 1999 the 2-3 dollars/barrel daily. Transit costs from the Iranian pipeline would be similar to the cost of the pipeline consortium that brought oil to Novorossiysk. The Tengiz oil field in Kazakhstan (6 billion barrels) can be the same size as the Kashagan oil field, whose production perhaps could peak at 700,000 barrels/day. The State Oil Company of Iran, NIOC, argued that crude oil from Kazakhstan would bring to the producers for sure $1 per barrel more, if transferred through Iran. The combination of fields and pipelines already created a lot of pressure on the Mediterranean markets, which absorbed 2 million barrels/day from the consortium of Caspian pipeline and the Baku-Ceyhan pipeline.

In 1999, the Oil Minister of Saudi Arabia Ali Naimi and the Foreign Minister Saud al-Faisal stated that the country had no need for oil companies programs in the upstream sector (research, development, production). The Morgan Stanley Securities was hired by al-Faisal to give advice on alternative investment options. Naimi said that "the share of oil in GDP has halved over the last 25 years and now stands at around 35%, although oil beholds 75% of government revenues and 80% of earnings from exports" ("Saudi Arabia: Government makes clear," 1999: p. 5). During the same period, the Japanese Arabian Oil Company was under pressure to invest in railway and petrochemical projects based on gas, in exchange for an assignment in the neutral zone. Saudi Arabia has over 3,200 recognised reservoirs, but 90 percent of its production comes from a small number of main fields, such as Abqaiq (1940), Ghawar (1948–1949), and Safaniya (1951). These oil fields use *waterdrive* technology to create high levels of oil flow. Nevertheless, water injection permanently bypasses large pouches of petrol. The impending depletion of oil fields such as Ghawar was going to be estimated with Enhanced Oil Recovery techniques, namely gas injection, thermal injection and chemical injection. What's more, the global depletion of reserves is accelerated by the technological revolution, consisted of 3-D seismic, horizontal drilling, multilateral well completions, subsea oil production techniques, which facilitate oil extraction without guaranteeing an increase in recoverable reserves. That same year (1999) Gamal Fahmi was sworn in as the new Minister of Petroleum of Egypt. Gamal Fahmi had previously moved from the state-owned EGPC Company to the position of vice president and CEO of the Midor consortium. At the same period, oil production was decreased in Egypt from 920,000 barrels/day to 770,000 barrels/day. However, the domestic gas demand amounted in 1999 to 1.6 billion cubic feet per day and increased by 8% annually. In the gas sector, the operating companies were BPAmoco and EGPC ("Egypt: is declining oil production," 1999). In 2002, Sameh Fahmi announced pipeline projects to transport Egyptian gas to Libya and Libyan oil to Egypt (620 km.). The path of the first section of the pipeline, which would channel 150,000 barrels/day of Libyan crude to refineries in the Mediterranean port of Alexandria, was fixed and cleaned from mines on both sides of the border. The Egyptian gas could be used either for electricity generation and water desalination or for export through North African network to Southern Europe.

The year 2000 marked for some the "zero time" to promote a wave of privatisations. In the late 20th century, important public Indian companies were the GALL natural gas company and the exploration, development, production company ONGC. In China, the National Petroleum Corporation in the upstream sector (exploration, development, production) and the Sinopen in the downstream sector (transportation, refining, and marketing) were active. On the contrary, in Europe and elsewhere the form of consortium was spreading, such as in the case of the participation of the Finnish Neste in Borealis, but also other types of partnerships, such as between Neste and the Baltic Rim, or the Austrian OMV, the International Petroleum Investment Corporation of Emirates or the Austrian Krems Chemie, etc.

5.4. Energy science and technology issues

There is general agreement that future demand for fuel will be, at least until 2025, mainly oriented to gasoline, although alternative fuels will become more competitive. A 'critical' period from 2025 to 2030 could bring forth one or more alternative fuels, which will be as viable as gasoline and diesel. Then, a period would come of decreasing consumption of gasoline and diesel, while favoring alternatives.

The World Energy Council predicted three different scenarios: a) 'green shock' as the development of energy demand for transport could be abruptly stopped, b) continuing into a confusing environment, c) control of the markets, while the demand for energy, for transport fuels, would increase by 175%. A common trend in these scenarios is that the cars will have decreasing importance compared with trucks and air transport: passenger cars are expected that in the future will absorb 30% of transport energy, compared to 50% in 2000.

There are three main types of engines: Steam engines, which require fuel that can be burned. Compression ignition engines that require fuels with a high level of ketene. Ignition engines with spark plugs, which require fuel with a high octane level. For spark ignition engines, there are only four alternative fuels to petrol. The best is the liquefied petroleum gas. Then come the compressed natural gas, methanol and ethanol. The main factors influencing the choice between alternative fuels are: The cheap price. The better performance and reduced exhaust gas emissions. The replacement of imported oil. Methanol cost in 2000 about 20% more than gasoline or diesel fuel. Also, methanol is highly toxic, may be absorbed by the skin and it is 100% miscible in water. Liquefied petroleum gas, when bulk, is cheaper than gasoline, but the vehicles should be converted. It also has advantages for a clean environment. Compressed natural gas requires compression devices or public compression stations. It is already used in New Zealand and elsewhere, and it is a reliable fuel for car and bus fleets. The most realistic, technologically, and in abundance alternative fuel for land transport seems to be the natural gas, according to World Energy Council, although it offers no cost advantages over gasoline, and achieves only a minimum reduction of exhaust gas emissions. For compression ignition engines the only alternative fuel to diesel is ketene fuels derived from vegetables, such as rapeseed oil. However, all fuels derived from vegetables have a low level of ketene, relatively high viscosity, and high melting points. Although they may be substituted for diesel fuel, they are operating much better if converted to esters with ethanol. Fuel from vegetables is 4-5 times more expensive than gasoline or diesel, without providing a comparative advantage for the environment. Alternative fuels with greater research interest are: gas, electricity, liquefied petroleum gas, methanol, ethanol, rapeseed methyl esters, and hydrogen.

Nowadays, competition with gas intensifies the excessive accumulation of fuel oil. For this reason, around 1997 Shell had achieved at Pernis refinery, Rotterdam, the construction of a unit for the combination of gasification of fuel oil with power generation. Similar programs began in Italy. Diesel, compressed natural gas, and liquefied petroleum gas have competitive cost, and lower, by 10-30%, emissions that affect global warming (compared to gasoline). In 1993, the proven gas reserves were 142 trillion cubic meters, while oil reserves were 152 trillion cubic meters of gas equivalent. In 2000, 300,000 vehicles in Argentina used compressed natural gas, 270,000 in Italy, 180,000 in the former USSR, 43,000 in New Zealand, 36,400 in Canada and 30,000 in the US. In Italy, the natural gas vehicle technology had already been developed by the mid-twentieth century, and in 2000 there was a chain of 260 petrol stations selling compressed natural gas.

In 1998, California had become a leader in the effort to increase the market share for 'zero emission vehicles'. Furthermore, Fiat produced the hybrid car Multipla. Toyota and Matsushita were pioneers in the development of nickel-metal hydride type batteries. In 1999, the RAV4EV was released, which was the first commercially produced electric vehicle.

Another technology was the fuel cells device that converts chemical fuel, such as methanol, into electricity. The most suitable fuel for fuel cells was hydrogen, according to the World Energy Council. After sufficient processing, methanol and natural gas may also be used. According to what might be expected, the alternative technology of the future will be the fuel cell. This is a combination of hydrogen and oxygen chemical reaction that produces electricity and steam. Fuel cells, by the summer of 1997, were used by Mercedes and Toyota.

At the same time, the dependency of the electricity production on coal was still in 2011 greater in the USA than the global average: 49% in the USA while 40.2% globally, in comparison with dependency on atomic energy 19.4% in the USA and 15.1% globally, on gas

20% in the USA and 19.7% globally, waterpower 10.1% in the USA and 18.5% globally, petroleum 1.6% in the USA, while 6.6% globally.

The Five Year Plan in China projected an increase in gas consumption from 81 billion m^3 in 2008 to 260 billion m^3, or 9.2 trillion cubic feet in 2015. Six huge LNG terminals were constructed in 2011 on the shores of the country, while gas was also channelled by the pipeline of Turkmenistan. Additionally, geologists and engineers were examining ways of developing domestic production. The development of the gas market had been benefited from the liberalisation of the natural gas prices in the USA in 1978.

From 2005 to 2011, natural gas derived from shale rocks increased from 4% to 23% in the USA. At this period, the global supply of liquefied natural gas increased by 58%, while shale gas reduced the ability of Russia, Venezuela, and Iran to exert influence based on their large reserves of natural gas. Except shale gas, there is gas in sandstones and methane in coal beds, as well. Gas is likely to be found frozen in hydrates in the continental shelves of the world, where it is estimated that there may be 1000 trillion cubic meters. The foreseen Golden Age of Gas, according to the International Energy Agency, will be marked by an increase of 1.8 trillion cubic meters in the annual global production by 2035, reaching thus the 5.1 trillion cubic meters. This scenario provides for a gas development 50% higher than the scenario that is used as a base, and requires the reduction of coal by 2019, and overcome of coal by gas in 2030, when covering ¼ of global energy consumption.

Nevertheless, there are severe environmental risks in using the fracking technology that drills down into the earth fluids at high pressure, in order to break open the 'impenetrable' shale rocks and release the gas, found far deeper the underground waters. Therefore, investigations had stopped in France, while in Poland, with the largest probable reserves of shale gas in Europe, companies open test wells with haste. Shell Company, which is positioned as closely tied with gas, did similar research in South African Karoo Basin and in China.

The damaging effects of extracting shale gas may not be greater than the existing in using coal. China reported 2,433 deaths of miners in 2010. In America, the same year, 23,000 premature deaths were estimated, while the number of diseases attributed to soot from burning coal in power plants became 20 times bigger. According to a study, if the damage to health and the environment were included in the price, then electricity from burning coal would be over 100% more expensive, while for gas the increase would be only 4%. Typically, coal produces twice the amount of carbon dioxide per kilowatt hour than gas. Europe seeks to cut emissions by over 80 percent until 2050. The pollution generated by sulphate due to coal, cools the planet as it shades the surface from sunlight.

Hydrogen is the ultimate fuel for the distant future, although the problem of storage of such light particles should be resolved. Hybrid vehicles can reduce pollutants consumption by 90% and achieve twice the distance, with the same consumption of gasoline (which means 50% reduction of the emission of carbon dioxide). The conversion of hydrogen in fuel acquires the optimum reliability with the fuel cells.

The troubles arise with the high transportation costs of hydrogen as bulk freight (ten times more expensive than crude). Regarding the technological innovations needed, the biggest problem is that with the storage methods available in 2000, only 3% of the total weight was hydrogen, while the rest was the weight of the tank. At Johns Hopkins, researchers studied the construction of a car that would burn ammonia (hydrogen carrier and free of carbon).

Chapter 6

Computer generations

The discovery of numbers, around 30000 BC, the cuneiform script and the sexagesimal notation, in 3000 BC, the invention of abacus, Gunter's Scale, and slide rule were significant benchmarks in the history of computing. The first computational machines operated by movable mechanical parts such as axles and pulleys, which all were manual; for example, the Antikythera mechanism was used for astronomical observations. The ancient and medieval scientists used also sundials, heavenly spheres, diopter, the astrolabon organon, the parallactic instrument and the mural quadrant.

The introduction of the alphabetical symbolism of unknowns and general parameters by François Viète (1540-1603), with its consequent separation between *logistica numerosa* and *logistica speciosa*, the invention of the slide rule by William Oughtred and Richard Delamain, the discoveries of the analytical geometry by Descartes and the calculus by Newton and Leibniz, the Mathematization of physical science by Galileo, the technological and scientific contributions of the Muslim astronomers and mathematicians, for instance the lunar eclipse computer of Jamshid-al-Kashi, were significant advances in the history of computing (Goldstine, 1993).

Analog computing devices appeared since the sixteenth and seventeenth centuries and contained graphs and logarithmic rules for shipping calculations. Decisive was the contribution of John Napier, who invented the logarithms and published them in 1614. With his second work, *Mirifici logarithmorum canonis constructio...*, published in 1619, Napier presented the principles of calculation with logarithmic tables.

The simplest automatic machines for adding-subtracting were invented independently by Wilhelm Schickard (1592-1635) and by Blaise Pascal (named *Pascaline*). Schickard's machine could also partially multiply and divide. Those calculating machines used geared wheels to perform additions and subtractions, and they were improved in 1671 by Leibniz. His device, now known as the *Leibniz wheel*, surpassed Pascaline that was limited to linear operations, permitting also multiplication and division. Based on the principle invented by Leibniz, the first digital mechanical calculator suitable for practical uses was the *Arithmometer* constructed by Charles Xavier Thomas de Colmar. In the era of the industrial revolution, people used computing instruments and information technology became extremely important. During the mid-nineteenth century, many simple ordinary mechanical integrators were produced. The idea of automating the computation of mathematical tables, such as logarithms, motivated the irascible genius of Charles Babbage. After 1850, the computers were being exposed in various international exhibitions. One of the most important steps for the development of computing was the foundation of Boolean logic, in 1854. Later, Boole's ideas supplied the basis for Claude Shannon's analysis of switching circuits.

The invention of the cash register by James Ritty and John Birch in 1879 satisfied the demand for practical utilization. The first electromechanical computing machine (since it did not contain electronic circuits) was constructed by Herman Hollerith, who also devised a data storage method with punched cards. In the beginning of the twentieth century, tables with various statistical, financial and banking accounts were widely disseminated. At the same time, computational techniques were further developed. One of those was the nomograms. In the year 1914, an international fair organized in Edinburgh was exclusively dedicated to computational technology.

The first generation of computers was based on vacuum tubes: "orange-hot filaments glowed in various computing machines from 1943 to 1959" (Crevier, 1993: p. 295). With the invention of the transistor by Bell Telephone Laboratories in 1947, the *second generation* of computers arrived. The transistor was a semiconductor equipment made of germanium or silicon that acted as a switch or amplifier. If current was channeled through a transistor, it passed to the one direction but not to the opposite. Miniaturization let computing performance speed up: A silicon integrated circuit (chip) may contain thousands or millions of transistors. Since the 1950s, many large companies involved with computers as either producers or, mostly, as users. From the 1960s, the space programs necessitated the construction of mini computers.

The *third generation* of computers, from 1971 to 1980, were constructed with tiny semiconductor circuits. They were performing 10^6 calculations in one second. The first tiny semiconductor integrated circuits were manufactured in 1965. The next year, 1966, at Stanford Research Institute the first modern modem was being produced. The modem device converted serial digital information into analog signals suitable for transmission over telephone lines and vice versa. The big bang was done in the 1970s, with the placement of basic units of a computer on a single integrated circuit, the *microprocessor*. The microprocessors are complex integrated circuits, which are able to process binary information from input devices and provide instructions for the operation of electronic devices. With the introduction of the microprocessor became possible to manufacture computers that occupy much lesser volume, consume less energy, have lower cost and increased capabilities.

> *On November 15, 1971, Intel announced its microprocessor with a bold advertisement proclaiming the arrival of "a new era of integrated electronics - a microprogrammable computer on a chip!" More than 5,000 people wrote to the address at the bottom of the advertisement, requesting more information - the most dramatic response to a product announcement Intel had ever experienced* (Berlin, 2005: p. 203).

The *fourth generation* was the Large Scale Integration technology, whereby 10,000 to 20,000 components were included on a single silicon chip. Subsequently, the Very Large Scale Integration technology was based on tiles with a surface of a few square millimeters, with thousands or millions of electronic components. By 1985, the microprocessor chips contained up to a quarter of a million elements. In the year 1985, the supercomputer Cray-2 was built that operated over $2x10^9$ operations per second. At the end of the twentieth century, they reached from $100x10^9$ to $200x10^9$ operations per second. Master computer of that period was the Fujitsu Numerical Wind Tunnel, which performed 170 billion of operations/sec.

The *fifth generation* of computers aimed at very high-speed calculations, in large data storage units, with the development of "smart" computer systems. Some of their most important applications emerged with the photo interpretations of digital images, the diagnoses of diseases, etc. Scientists envisioned developing the findings of information technology and, in particular, of artificial intelligence, into fuzzy-logic computers, built into microchips, capable of decisions under uncertainty. The challenge now was to build a *quantum computer*, which would have allowed the *factorization* of any large integer, in a few seconds.

6.1. Charles Babbage and the development
of computational technology

The Englishman Charles Babbage (1792-1871), Lucasian Professor of Mathematics at the University of Cambridge, was the first computer scientist. He wanted to build a machine that incorporates the disciplined structure of social work in the computer. Charles Babbage designed the digital general-purpose computer and constructed the *Difference Engine* in 1823, funded by the British Government, and the *Analytical Engine* in 1833. Those plans were extending far ahead from the needs of his times and the technical infrastructure for their implementation. The use of punched cards, chains, subsystems and the logical structure of modern computer spring all from Babbage.

The Difference Engine was built for the automatic production of mathematical tables (such as logarithmic tables, tide tables, and astronomical tables). In 1827, Babbage had published a table of logarithms that started from the number 1 and reached to the 108,000. In 1834 he founded the Statistical Society of London. When Babbage brought the Difference Engine in the house of Lord Byron, the daughter of Byron, Ada Augusta Byron, was impressed by its capabilities, with the result to work for years together with Babbage in programming the machine. However, they did not experience success. Babbage also worked on inventions related to the lights of lighthouses, the signals for the Greenwich Time, the ophthalmoscope, etc.

The Babbage Principle

The British nineteenth century was an unparalleled creative era. The engineers of London's Lambeth Borough worked for the nobles of the West End, looking for the approval of their projects by the Royal Society, the Astronomical Society and the Royal Institution. Babbage and Darwin lived in the fashionable Marylebone. Huge working-class neighborhoods were expanding to the Northeast, as Schaffer (1994) notes. On the quest of exemplifying the industrial revolution, Karl Marx (2011) reiterated lots of enlightening quotes from Babbage's writings: "When each process has been reduced to the use of some simple tool, the union of all these tools, actuated by one moving power, constitutes a machine," according to Babbage

(1832: p. 136). "It has been estimated, roughly, that the first individual of a newly-invented machine will cost about five times as much as the construction of the second," as Babbage (1832: pp. 211-12) continues.

> *The improvement which took place not long ago in frames for making patent-net was so great, that a machine, in good repair, which had cost £1200, sold a few years after for £60. During the great speculations in that trade, the improvements succeeded each other so rapidly, that machines which had never been finished were abandoned in the hands of their makers, because new improvements had superseded their utility* (Babbage, 1832: p. 233).

The economic aspects of industrial life, being swiftly modified, should be comparatively calculated. As the careful reader concluded: "In these stormy, go-ahead times, therefore, the tulle manufacturers soon extended the working day, by means of double sets of hands, from the original 8 hours to 24" (Marx, 2011: p. 442).

> *When (from the peculiar nature of the produce of each manufactory), the number of processes into which it is most advantageous to divide it is ascertained, as well as the number of individuals to be employed, then all other manufactories which do not employ a direct multiple of this number will produce the article at a greater cost... Hence arises one of the causes of the great size of manufacturing establishments (...)* (Babbage, 1832: pp. 172-73)

Although the mathematician and engineer Babbage viewed, according to Marx, the mechanical industry from the aspect of manufacture alone, his opinion was considered as very significant: "Babbage estimates that in Java the spinning labour alone adds 117% to the value of the cotton. At the same period (1832) the total value added to the cotton by machinery and labour in the fine-spinning industry, amounted to about 33% of the value of the cotton" (Marx, 2011: p. 427). The whole idea of the division of labor was attracting the social researcher:

*The master manufacturer, by dividing the work to be
executed into different processes, each requiring different degrees of skill or of force, can purchase exactly
that precise quantity of both which is necessary for
each process; whereas, if the whole work were executed
by one workman, that person must possess sufficient
skill to perform the most difficult, and sufficient
strength to execute the most laborious of the operations into which the article is divided (...)* (Babbage,
1832: pp. 137-138).

Accordingly, the design of a machine is typically shaped by the
needs and aspirations of the social and technological environment,
claim historians as Miller (1990). The construction of engines that
perform mathematical calculations should be approached in relev-
ance to the social context. Computation, during the nineteenth
century, was inherent in the most essential aspects of English culture, which inspired and made timely the work of Babbage. Already
by the early eighteenth century, intensive industrialization was
accompanied by extensive interests in the social and economic
consequences of technological innovations. This interweaving of
technological and social transformation was more vigorous in
Britain than anywhere else. With the industrial revolution, all the
variety of the facets of life that could be measured and manipulated,
acquired increased importance.

Babbage, specifically, showed how much the improvement of the
design of machines contributes to the advancement of society and of
the state structure. He supported, therefore, the *division of labor* into
the areas of skill and power, because such a division would enable the
employer to hire the exact amount of competency and strength he
needs anytime, according to the circumstances. By this division, the
worker himself will no longer be obliged to possess simultaneously
sufficient skill for the perplexed and adequate power for the most
arduous tasks. Any art can be divided into components of varying
complexity. The adjustment of the abilities of each worker to special
mission requirements and the differentiated payroll, based on skill
level, may really yield, according to Charles Babbage, economic advantages. This innovative proposal for the division of labor, known as
the *Babbage principle,* was similar to the ideas of Adam Smith, who
avoided burdening the craftsman both with the more elaborate mis-
sions and the drudgery of the low specialized roles.

Babbage and De Prony

Since 1791, the French Government, wishing to establish a fair system of property taxation and to disseminate the use of the metric system, assigned to the mathematician and engineer Gaspard De Prony the construction of up-to-date maps of France and tables of land registry (tables du cadastre), which necessitated the production of large logarithmic and trigonometric tables. The French researcher divided the work of producing extended mathematics tables into three levels: a) The overall strategy of the project and the most useful mathematical formulas of the five-six most famous French mathematicians, including Adrien Legendre and Lazare Carnot. Generic types served the necessary mathematical calculations. b) The practical numerical work conducted by seven–eight people, capable in mathematics, which transformed the formulas in numbers, giving values to the variables. c) The third level was covered by sixty to eighty human computers, which carried out the calculations with the "method of differences," requiring only additions and subtractions, and, finally, compiled the tables (Miller, 1990).

Babbage was influenced by De Prony's activities. In his book *On the Economy of Machinery and Manufactures* (1832), renowned for his extensive knowledge of industrial processes, Babbage, with watch in hand, discovered the enterprise research, namely the scientific study of industrial production, and examined the accumulation and the coordination of power. By the use of the wheel, for example, power is stored. Further, with the springs and wheels of watches, power is built up and maintained.

One of the most common and useful applications of machines is the extension of the time of action of forces, the gradual release of energy, as with watches, and the stop-restart of power-source operation, as it is done with the automatic control device (governor) and the thermostat. In watches, the tension of the spring increases while it is stretching, and then gradually and continually decreases, as it is slowly rewound. In short, the power of the past is activated in the present.

The program developed by Babbage

In July 1822, Babbage wrote an open letter to the president of the Royal Society, Sir Humphrey Davy, describing his plan for calculating and printing mathematical tables through a machine. In June 1823, Babbage obtained funding of £1,500 by the Chancellor of the Exchequer and started next month. In 1829, he was sponsored again, by the Duke of Wellington. The Difference Engine was planned to calculate up to twenty decimal digits and six orders of differences. Thus, with additions and subtractions, calculated with the method of differences, a polynomial was represented in a sort of a tabular form. The prototype built in 1822 calculated only two orders of differences. For the construction, he used subtle mechanisms of tin, steel, and brass.

In Babbage's engines, the operation that represents with the greatest clarity the phenomenon of temporal relationship is the *mechanical method of transferring a digit* to the next column. An addition with the first machine lasted nine seconds. Every time that a wheel was passing from the ninth digit to the digit zero, a tooth placed between nine and zero, *was overturning a certain lever.* The result was to rotate an arm that carried the designation to the next level upwards. The arms were executing successive transfers, i.e. *carry propagation,* by being placed spiraling around an axis. Babbage paralleled his engine with the mechanism of memory. Nevertheless, the successive transfers were spending too much time, which annoyed him. He also thought to promote the prognostic features of the machine, but he did not realize his plan. The storage operation of his engine consisted in making a recording of a certain change in the arrangement of levers and wheels, or in the overall structure, or system state. The peculiarity, in comparison with other machines, it was that computers stored rather *information* than energy or power. A difference from modern computers was that the programming of the machine was accomplished by the *physical reconstruction* of the system rather than with a standard set of instructions.

In late 1830, when they had to move their workshop, Babbage and his assistant, Clement, quarreled upon disagreements and financial requirements, which led to the termination of the program. The new machine, the analytical engine, which Babbage was planning, involved 40 decimal digits, dealt with negative numbers and performed multiplications. That engine would perform calculations of any function. The Government, however, did not want to finance a new engine, while the old one had not been completed, though it cost £17,000 pounds. In November 1842, Babbage received a decisively negative response, so his work remained at engineering drawings level.

Babbage was unable to complete his plans, ignoring that no information transfer is possible without losses, by analogy with the principle of entropy. In the positive elements of that project, one may include the good attachment to the practical activity, which is evidenced by the blacksmith that he kept at his workshop. Babbage constructed numerous tools, such as tin wheel gears, for the Difference Engine, and designed lathes and machine tools. Joseph Whitworth, the janitor in Babbage's laboratory, introduced the first series of standard screw threads. Around 1846, at the end of his work on the analytical engine, for approximately 15 months, Babbage worked on another machine, the Difference Engine No. 2. In the early 1850s, Babbage dealt with the development of a mechanical system of symbols (mechanical notation). In the mid-1850s, Babbage returned to the Analytical Engine and tried to simplify it. Shortly before he died, he had almost completed a test model with two storage axes and a forecast - transfer technique (anticipating - carriage).

The Method of Differences

The work of De Prony, in its second stage, involved the conversion of formulas into numbers, by *assigning to the variable values,* which were different according to certain successive distances. The same principle was applied in Babbage's Difference Engine. By examining the series of the squares of numbers 1, 4, 9, 16, 25, 36, 49, 64 etc., and subtracting each and every one of these numbers from his next, we take a new series, which we call *Series of the First Differences,* i.e. composed of the numbers 3, 5, 7, 9, 11, 13, 15 etc. Furthermore, by subtracting from each one of these differences, the preceding one, we take the *Second Difference,* that is always fixed and equal to the number 2. The successive values of a polynomial function are given by the aforementioned method of differences. The sequential and concise character of the calculations performed in this way was revealing a completely new element, which we call today *pipelined operation;* in each stage of which, a new result is extracted, through a direct information channel. Horizontal wheels rotating upon vertical axes store the numbers. To every digit corresponds a wheel that represents the units, the tens, the hundreds etc.

Programming the Analytical Engine

The *Analytical Engine* was designed to use two extremely deep and innovative ideas, which constitute the entire groundwork of computer science: a) the functions were fully programmable, and b) the programs could contain conditional branching. The Analytical Engine had three main parts: the Mill, namely the arithmetic processing unit, the Store, i.e. a data memory module, and a unit for synchronization and control of operations, to which Babbage did not assign a name (Haugeland, 1989). The symbols that the system handled were numerical with sign and the field of the game was substantially unfolded in the memory unit, i.e. the Store.

The Mill could perform the four common operations, using for the operators and the results any position indicated by the Store. Giving different definitions one may transform the analytical engine to any desirable automatic formal system. This remarkable activity, the realization of a particular automatic system, described properly in a general-purpose system, is called *programming*. This fruitful idea was a discovery of Babbage. The second great invention was the *conditional branching*: a transition instruction to another part of the program, depending on the result of control. Babbage's programs were essentially branching directories.

The representations of the numbers allowed for forty decimal (not binary) digits and a sign. Each digit was represented by rotating a toothed wheel, sized approximately as a little pretzel. The forty wheels representing the digits of a number were arranged in a bronze axle with a height of three meters approximately. The *Mill* consisted of at least one hundred similar columns, ordered in a circular shape, with a diameter slightly greater than one and a half meter. The *Store* contained another hundred columns, placed in a double row, extending out from the side of the mill. All these, of course, included a series of levers and inhibitors, gears, and clutches, for determining and monitoring the appropriate movements, as well as with the necessary precautions from friction and reciprocating motions.

The Analytical Engine programs were not embedded in the *Store* but they were encoded in punched cards, hanging in a row of cords. With this system, the machine could return and repeat a group of commands over and over again, which is necessary for the implementation of the loops. This basic layout was inspired from a system used in the automatic production of embroidered designs woven by the Jacquard looms and this fact inspired to Ada Augusta the famous simile that the analytical engine weaves algebraic patterns. Similarly, the intellect weaves symbolic shapes.

The punched cards performed control functions and were divided into four types: *number cards, variable cards, combination cards* and *operation cards*. The most general operations were controlled by the variable cards, defining the governance of memory, and the operations cards, which guided the Mill to perform addition, subtraction, multiplication, division and certain other mathematical operations. The variables can be regarded as a repository of numbers, accumulated by the *Mill,* and, following the instructions that are transferred to the machine through the cards, passed alternately from the *Mill* to the *Store* and from the *Store* to the *Mill.*

The intelligence in the work of Babbage

The term intelligence refers both to the signals received from the external environment, as well as to the ability to interpret these signals. At the beginning of the nineteenth century, in Britain, the term *intelligence* included the rising social surveillance system and the emerging *mechanization* of natural philosophies of mind. According to Simon Schaffer (1994), the addressing of machines as intelligent was marching together with the concealment of the contribution of the labor force in their production, while, at the same historical period, the critical problems of the status of technical skill in the social pyramid and of intellectual property were raised widely. Of course, the *Difference Engine* was based on a mathematical principle: that the successive differences of polynomial values were finally constant. Thus, they discovered that they could construct, in this way, tables of values, by adding and subtracting the predefined constants.

The *Analytical Engine* was, as we have seen, a more ambitious program, which, although it had no tangible material results, gave the basic architectural guidelines for the development of computational science. Whereas the *Difference Engine* was invented to improve the accuracy of mathematical calculations, complementing thus the mind, the *Analytical Engine* had a more flexible design, which permitted the analysis of various types of problems.

> *Babbage's Analytical Engine marked a cognitive shift*
> *from engines of accuracy to engines of abstract, flexi-*
> *ble analysis (precursors to the modern digital comput-*

er). This distinction between accuracy and flexibility
in Babbage's calculating engines is also relevant to
properties of contemporary technology (Mather, 2006:
p. 239).

Within the broad scientific and technological advances of the era, the achievements of Babbage confirm once again the increasing importance of mathematical literacy and the impact of technological innovations and inventions. The craftsman and the scientist, successively in the history of science, as has happened in the Renaissance, with Leonardo, and in late Antiquity, with Archimedes, occupy a common breeding ground and their activities are constantly feeding each other.

Intelligence, however, acquires hereafter a special importance, completely autonomous and innovative, as well as economic value. The epistemological problem of the *intelligence* attracts today, remarkable importance for the historical study of the transition from the industrial era, which was characterized by the expansion of *machinery mass construction*, into a new era, dominated now by the design and construction of machines *by the machines themselves.*

6.2. The emergence of the computer industry

Apart from the invention of the first digital arithmetic machines, significant advances in the early development of computing were the conception of the Three-Bodies problem by Isaac Newton, the construction of analog measuring devices such as integrators and planimeters, the interferometer, the echelon spectroscope and the harmonic analyzer of Albert Abraham Michelson, the numerous researches on ballistics, and the differential analyzer of Vannevar Bush (Goldstine, 1993).

Between 1937 and 1945, the German engineer Konrad Zuse constructed four computers, called Z1-4. In America, the Atanasoff-Berry computer was produced by 1942 - with binary function, but not completely programmable. In 1944, the Mark I and the ENIAC (Electronic Numerical Integrator and Computer) were developed. In England, the Colossus was built in 1943, the EDSAC (Electronic Delay Storage Automatic Calculator) in 1949, and the UNIVAC (Universal Automatic Computer) in 1951 (Wieland, 2013).

Around 1948-49, Claude Shannon and many other scientists contributed to the foundations of computer science:

*In 1948, John Bardeen, Walter Brattain and William
Shockley of the Bell Telephone Laboratories published
the results of their work on solid-state electronic devices, which began just after the war in 1945. This paper described the invention of the point-contact transistor, which has almost completely displaced the
valve as the active component in electronic circuits.
For this work, the three shared a Nobel Prize in 1956.
Also in 1948, Claude Shannon of Bell Labs developed
his mathematical theory of communication (also
called information theory), the foundation of our understanding of digital transmission. In 1949, Forrester
invented magnetic-core storage for the Whirlwind,
which became the standard internal memory for large-
scale digital computers until semiconductor memory was
introduced in the mid-1960s* (Ohlman, 1990: p. 703)

Alan Turing, an English mathematician, equally brilliant with
Babbage, became famous because, during and after World War II,
contributed decisively to the ground-breaking British research into
computers. He created the first mathematically exact computational theory, which included some impressive discoveries and invented a new basic construction of computer architecture. According to him, a mechanical process, such as the computer programming, is defined on the basis of rule-following (Hodges, 2012).

The computer ENIAC (1946), designed at the University of Pennsylvania, was constructed with 2x104 vacuum tubes. The ENIAC was
enormously voluminous, consumed a lot of energy and performed
a small number of calculations compared to later computers. However, ENIAC gave an enormous boost, being used for compiling
tables of reckoning artillery firing and later for calculations relating
to atomic energy and weather forecast.

John von Neumann emigrated to the U.S.A. from Hungary. He developed the first practical general-purpose computers, after 1946.
He conceived the idea that the instructions for data processing
should be recorded in the computer's memory, along with the
processed data. Therefore, in 1952, the Electronic Discrete Variable
Computer (EDVAC) was built in Cambridge, the first computer with
a stored program, a concept that was the basis for the creation of
the *software*. Afterwards, the necessary *programming languages*
began to disseminate.

John Von Neumann's contribution was significant in the domains of mathematics, economics, scientific method and artificial intelligence. According to his study, while the nervous system, i.e. a natural automaton, being much more efficient in energy terms, has a million times as many components as the machines have, yet, each component of the artificial automata called computers was then about 5 thousand times faster than a neuron. Actually, Von Neumann introduced the theory of automata, invented the game theory, contributed to hydrodynamics with calculations of shockwave propagation, and offered a better computer architecture in 1945. He emphasized also on the repetitive basis of memory, revealing a peculiar field of theorizing through Quantum Mechanics (Birkhoff and Von Neumann, 1936).

6.3. Algorithms and their power

Calculus evolved from two ancient problems: Attempting to find the tangent (the slope) to a curve, and trying to estimate the area under a curve. From Archimedes to Newton, a time-interval of more than nineteen centuries was needed, until the *necessary* and sufficient conditions prepared the realisation of the qualitative leap to the differential calculus and the integral calculus respectively.

In the beginning of the twentieth century, tables with various statistical, financial and banking data were widely disseminated. At the same time, computational techniques were further developed. One of those was the nomograms (since 1884). The turn to generalisation after the conquering of calculus, provoked a proliferation of theoretical, semantic and experimental approaches (Russell-Whitehead, Zermelo-Fraenkel, David Hilbert, J. von Neumann, Kurt Gödel etc.), which gave birth to innovations such as the computer science and the subtle processing of data structures, in the form of arrays, records, lists, stacks, queues, tables etc. Perhaps, some of the most important analytical contributions to this evolution are the set theory and the theory of genericity, which includes the parametrisation of classes by types and the constraining of types.

In computer science, there are also significant conventional separations such as between interface and implementation (information hiding) and between command (modification) and query (access) (Meyer, 2009). The dissemination of the algorithms and the computer science necessitated also combating security threats through encryption, by means of primality and factoring. Moreover, the specific advantage of quantum algorithms is their exponential power, which is featured by the Big-O notation (Papadimitriou et al. 2010). Quantum machine-learning aims at the improvement of patterns "derived ('learned') from data with the goal to make sense of previously unknown inputs" (Schuld et al. 2015).

Chaos theory occurred naturally from the ground-breaking difference between linear and non-linear causality. Possible better and more accurate definitions of the chaos issue could be "nonlinear dynamical systems theory" and "complex systems theory". A special sub-category of the aforementioned is the catastrophe theory.

In the 1980s and early 1990s, at Los Alamos National Laboratory and Santa Fe Institute, Artificial Life experiments by Christopher Langton, Thomas Ray, Christoph Adami etc. formulated the nonlinear dynamical systems theory, one of the most innovative theory-practice relay systems in contemporary science and technology (Johnston, 2002).

In the same way, autonomous mobile robotics -where kinematics, locomotion (legs and wheels) and perception are important- use modern mathematical concepts, such as the Poincaré section, the Poincaré map, the monodromy matrix and the eigenvalue analysis, for computing. Moreover, the theorising of chaotic attractors and dynamic systems proposes modern research projects under the general categorization "virtual environments" (Stanney, 2002), including cyborgs and androids.

The classical notions of cause and effect are replaced by concepts involving control in the engineering sense, bifurcation, energy, and turbulence (Guastello, 1995). The perception of dimensionality also changes, as happened after the invention of *fractal geometry* by Benoit Mandelbrot (Horvitz, 2002). In physics, especially in modern optics, the study of non-linear processes is indispensable, as with the experiments with lasers in the SCSS Test Accelerator (Japan), FLASH (Hamburg) and LCLS (Stanford) (Berraha et al. 2010).

Likewise, the deconstruction of the traditionally discrete environment of the rational numbers, by the immanent character of the real numbers and the limits, opened new paths to research. One of the most intriguing applications of the differential equations is the study of the rate of growth of population, especially in the USA, with the implementation of computing models such as the prey-predator system of equations (and the excessively important equilibrium solutions they may have) (Véron, 2000).

Moreover, the visual methods became more important, as for instance with the use of graphs for the study of derivatives, convex and concave functions, and with the corresponding application of convex combinations of matrices for the creation of visual art. The applications of linear algebra, namely of the vector norms, range from data mining to the deciphering of handwritten numbers.

6.4. Innovation in the age of Technoscience

Innovation is interconnected with mechanical engineering, for instance, with the expansion of transportation. After 1888, the German Empire was investing in the Anatolian Railway (Quataert, 1977), connecting it with the Bagdad Railway. That project encompassed linkages between innovation, profits, stocks, bonds, troop mobilization, central control, and political-diplomatic interests. In fact, the Anatolian Railway was competing efficiently and substituting caravans and camels.

Innovations were historically related to the long cycles of economic development, firstly, the steam engine and textile industries in the end of the eighteenth century; secondly, the railways, the mechanical engineering, the iron and steel industries; thirdly, the advent of the electric power, the internal combustion engine and the chemical industry. Five waves of key transformations were involved, according to Ayres (1990): The shift from charcoal and water-power to coal for the purposes of iron making (1770-1800), the introduction of steam engines, canals, and mechanised cotton spinning. The application of steam power in railway and steamboats (1830-1850). Steel making and mechanised manufacture on illumination, telephones, electrification and internal combustion engine (1860-1900). The introduction of synthetic materials and electronics (1930-50). The combination of computers and telecommunication (since 1980).

Over and above that, business historians such as Alfred D. Chandler Jr. and James W. Cortada, suggested that the information age had been prepared for more than 300 years, with highway construction, postal system, copyright laws, newspapers, books, pamphlets, broadsides etc. From this perspective then, oil industry and global transportation were indispensable factors of the information revolution. After 1865, the emergence of new forms of business, especially in the USA, created a period of rapid and widespread economic growth. Since 1900, the automobile became the bearing vehicle and the specific difference between consumption styles in the beginnings of that metamorphosis. The oil industry remodelled the economy, built cities, attracted investment, created jobs, funded educational projects, redefined government's role in industry, and generated unprecedented wealth.

Oil companies were networked with complementary industries, such as lumber, railroads, ports, machine shops, oil supply, and equipment (Olien and Olien, 2002). The role of *entrepreneurs* and *company builders* was critical in this transformation, along with organisations such as the Rockefeller Foundation, whose activities spans through numerous realms of research interests, from oil to biomedicine and number theory.

There was also a combination of oil technologies with the quasi-urbanising technologies (Kline, 2000) of telegraph, telephone, phonograph, home appliances, computers and motion pictures. The clinching turn was taken in the 1920s with the introduction of the vacuum tube and the electronics, "a new power source," which disseminated the telephone lines and diffused the communication networks as never before, all over the world. The vacuum tube could be used to rectify, to amplify, to generate (for instance, the radio transmitter), to control, to transform light into electric current, and to transform electric current into light. The successor of the vacuum tube was the transistor, more power-saving than the vacuum tube, more reliable, smaller, more effective, resilient etc.

The value of an innovation is relative to the extent of its implementation. There are two types of discourse over innovations, based on: a) the developer or source that control research, development, testing, manufacturing, packaging, dissemination, etc., b) the users who express their awareness, selection, adoption, implementation, routinisation, etc. (Klein and Sorra, 1996).

Only after the second half of the twentieth century, small enterprises emerged again as considerable agents of innovation. In the twenty-first century, risk-taking, competitiveness, freedom to fail, the venture capital, the technical infrastructure of prototyping new devices and outsourcing components, are considered as prominent parts of the innovation machine. This is why the Silicon Valley, with its 300,000 top scientists, has been characterised as the ultimate cluster phenomenon. The dynamism and the growth of the clusters are interconnected with skilled workers and technological innovation. Small and Medium Scale Clusters develop themselves from Latin America to Asia and Africa, according to the following types: (a) diversified industrial cluster; (b) the subcontractor cluster; (c) the market town-distributive cluster and (d) the specialised petty commodity cluster (Uzor, 2004: p. 7).

Different groups intervene in the course of innovation, alternating and transforming the significance and the margins of scientific, technological and industrial knowledge. Thus, innovation may require various types of knowledge, i.e. related to natural world, design practice, experiment, final product, and knowledge per se (Faulkner, 1994).

Last but not least, ecology and sustainable energy appear to be one of the most interesting areas for innovative enterprises. For ecological economics key concept is the ecosystem, the bio-geo-chemical system, industrial metabolism (Ayres, 1997) and industrial ecology (Weston and Ruth, 1997). In theories such as Ayres' (1997), the emphasis is on the competitive free market, while companies should make sure to recycle all the waste of industrial capitalism.

6.5. Work, technology, and Human-Machine Interaction

The introduction of technologies in society is constantly question-ing the human component in working relations while transforming the human conceptions of work. Already by the early eighteenth century, extensive interests in the social and economic conse-quences of technological innovations accompanied intensive indu-strialization. In the early stages, the computers facilitated the statis-tics and the census. The precursor of IBM, the Computing-Tabulating-Recording Company was born in 1911, after the inven-tion of the punched-card tabulating device by Henry Hollerith, at the request of the U.S. Bureau of Census to process data collected for the census of 1890. Thenceforth, not only the collection and the registration of the data was better organised, but also the processes of control, correction, and validation tweaked.

From the outset, many women were working in mass production and computers, as with the ENIAC (1946). The reorganisation of the working conditions was deepened after the introduction of the transistors, which led to the invention of the software (with sophis-ticated programming languages such as FORTRAN) and to a tenfold increase in the number of electronics engineers and technicians working for IBM (Chandler, 2005). In the late 1970s, with the diffu-sion of personal computers, the writing activity overlapped with composition, the audio-visual processes enhanced in educational practice etc. Simultaneously, commercialisation is widely intertwined with digital technologies, strengthening their central role in the society. After the introduction of the radio and the television, a turning point in the evolution of global electronics was the competition between RCA, Philips, Matsushita and Sony for the commercial exploitation of the videocassette recorder.

Until the end of the last century, the new disc technology produced the audio compact disc, the CD-ROM and the digital video disc (DVD). New technologies favoured the creation of massive businesses that needed and gave rise to superstructures of professional managers: engineers, accountants, and supervisors, extending to autonomous divisions, general office, multidivisional structures, as Alfred Chandler (1962) has shown.

However, computer technologies challenge more openly the traditional pyramid management hierarchy, permitting flatter structures, instantaneous communication, defragmented and thus integrated projects (McConnell, 1996). At the same time, the introduction of the Automated Teller Machines changed consumption and employment conditions, and made commercial banks more competitive, by offering additional new products and services.

Computer technology changes the quality and the quantity of jobs, rearranges the distribution and the composition of labour, by increasing effectiveness, creating new markets and new forms of organisations. Telecommuting and telework is one of these options, facilitated by portable computers and smartphones. "The computer industry is hugely splintered. Some firms sell components (Intel, AMD); some, software (Microsoft, SAP); some, services (IBM, EDS); some, hardware (Dell, Apple). There's overlap, but not much" (Samuelson, 2006).

The dissemination of small and bigger enterprises, which function as customers-network for the first-mover industrial leaders, was a direct result of the electronics and computers industry. In Europe, however, the ICT industry was not competitive against the Japanese industry and the USA.

Human work and programmable automation technologies

The rapid technological changes in microelectronics and communications create a requirement not only for different competence but also for ongoing renewal of the workers' present competence. The production of services and goods, while interdependent, becomes increasingly knowledge-intensive. This is why a shared understanding of the social function and conditions of work is the prerequisite of collective competence, such as navigating *a large vessel in and out of narrow waters* (Hutchin, 1993).

According to the official categorisation, at the end of the last century, there were at least five main programmable automation technologies: Computer-aided design (CAD), industrial robots, numerically controlled (NC) machine tools, flexible manufacturing systems (FMSs) and computer-integrated manufacturing (CIM). The roles assigned to the programmable automation referred "to facilitate information flow, coordinate factory operations, and increase efficiency and flexibility" (U.S. Congress Office of Technology Assessment, Computerized Manufacturing Automation, 1984). Moreover, they promised the optimisation of the management control over operations, something that is not always legitimate, because too much control can also inhibit creative participation and degrade the working environment.

The predictable changes in the labour market referred to an increase in the demand for engineers and computer scientists, technicians, and mechanics, repairers, and installers, alongside with an increase in the demand for clerical personnel, upper-level managers and technical sales and service personnel. On the contrary, the demand for craft-workers (excluding mechanics), operatives, and labourers was predicted to decrease. Although programmable automation displaced some jobs, it made firms more competitive and helped them employ more people. On the other hand, in 2003, the so-called "jobless recovery" in the U.S.A. related to the investments made by companies in computers in the 1990s.

Shared, global knowledge, and complex virtual systems

The development of cybernetics was based on the concept of *feedback*, which the mathematician Norbert Wiener defined as a method of controlling a system by evaluating the data of its past operations (Conway and Siegelman, 2005; Mayr, 1970; 1971). The first movers of the information industry created integrated learning bases that include barriers against their competitors. In 1970, a leading designer of the mainframe computer System 360, Gene Amdahl, left IBM and turned to Japan, "providing Japan's fledgling computer industry with state-of-the-art technology" (Chandler, 2005: p. xii). The following years, the four major Japanese computing companies Fujitsu, Toshiba, NEC, and Hitachi expanded globally. However, IBM revolutionized information technology one more time in the 1980s with the introduction of the personal computer.

A computerized manufacturing society requires complex networks of producers, agents, and consumers that combine and cooperate in organized systems (Mumford, 1966). Although the U.S. Department of Defense initially developed the Internet, banks and insurance companies played a significant role in the expansion of computers (Volti, 2014). Luddite practices against computers and technology in general originate from the disturbances in the labor market and they are understandable as feedback trials for negotiation.

Furthermore, a characteristic attribute of modern technologies is their interactivity: "*Something is interactive* if and only if it (1) is responsive, (2) does not completely control, (3) is not completely controlled, and (4) does not respond in a completely random fashion" (Smuts, 2009: p. 65). Video games and historical computer simulations share the two significant properties of Babbage's Proto-Computers, namely accuracy and flexibility (Mather, 2006). The use of computers today expands in various automatic applications, for example, vehicle steering. Technology becomes increasingly elaborate with the exploitation of modern scientific discoveries, such as fuzzy logic (Wieland, 2013).

6.6. The Human-Machine Interaction in future smart societies

The global perspectives for the future advancements in ICT include automatic control, electric autonomous vehicles, automated trucks, intelligent transportation systems, modelling and simulation, the transition from big data to smart data, holistic electro-mobility simulation, automotive cyber-physical systems, monitoring infrastructure, wearable technology, bio-nanosensors for healthcare, intelligent living rooms etc.

The ubiquitous reality of Human-Machine interaction calls for critically smart approaches, which surmount technological difficulty and tend to endorse an invisible, transparent, human-centred design approach. An appropriate example is smart cities in the domain of energy efficiency with the implementation of automatically adjustable systems and networks. Today, however, we realise that the cities are not sustainable, as by 2050 the urban population, everywhere in the world, will exceed the rural one.

For this reason, research programs explore the city perspectives, as in Singapore.[7] Endeavours to enhance services, to provide better connections, increase cooperation and preserve sustainability may offer smart instances (Kesrouani, 2015). The use of telemetry technology made possible the intelligent transformation of everyday life machines such as the automobiles. A similar transformation runs in the fields of digital technologies, fibre optics, telecommunications, sensors, and databases. Therefore, it allows urban research to imagine cities with driverless cars and free spaces, as Singapore needs. Smart visions are present also in Myanmar, seeking to transform the megacity Yangon to a smart city (Dale and Kyle, 2015).

The future home will imply Wi-Fi operated home appliances, home automation devices; and automatically adjusted thermostats. Inventions may enable every viewer to construct his own movie, insert his favourite actors and create holographic films. Artificial intelligence (AI) finds applications in the cinema, as with Skynet, a mental experiment with the power of rationality. Another instance

---

[7] (FCL) Future Cities Laboratory Singapore, Singapore-ETH Centre. http://www.fcl.ethz.ch/

of AI applications is the Genetic Lifeform and Disk Operating System (GLaDOS).

A key parameter in the building of smart environments is the estimation of the probable functionality of any innovation introduced. In a city, this estimation of probability is difficult and sometimes impossible, while it is easier in the country. Accordingly, a key feature in a future smart society is access and distribution of knowledge. In other words, smart societies are knowledge societies (Mansell and When, 1998). That is why education is an important part of the future smart society, for technology demonstration and knowledge production, as with the theatrical robots in Japan and Taiwan (Lin, 2015). Another example, the Trash Track project in MIT, aims at tagging and tracing products across their entire lifecycle.[8]

Inventing a smart innovative century

Technology becomes portable, wearable, pervasive, interfusing, reliable, adaptable, flexible, and increasingly personalised. The tailoring of the technological applications, however, brings about the problem of the embroilment between user and tool, when the tools, for instance, become part of the mental apparatus, or ubiquitous under the flying thumbs of the person (Clark, 2003). Moreover, proposals, such as the cyborgs (Cybernetic Organisms), that is to say, man-machine hybrids "destined" to explore the universe, open up interesting ethical problems (Yi, 2010; Pinsky, 2003).

Further ethical and legal issues arise from incidents of robots committing murders against their colleagues. The first incident was a homicide in Kawasaki Heavy Industries plant in 1981 (Hallevy, 2013). For such reasons, we should mention the demand for emotionally intelligent computers, learning companions, which will be sensing when they must intervene, understanding the origin of the students' mistakes etc. (Matthews et al. 2007). The ability to imagine integrated, intelligent, flexible, efficient, sustainable clusters of future production, mobility, education, residence, and leisure is an

---

[8] MIT, SENSEable City Laboratory, Trash Track (2009).
http://senseable.mit.edu/trashtrack/

essential presupposition for innovation and creativity. ICT is a key resource for the planning, the research and the monitoring of the networked elements of smart societies.

ICT are also important tools for harnessing development because the developing countries can no longer seek advantages on cheap industrial labour, but they need to applicate knowledge. The main indicators of developing countries' participation in "knowledge societies" are infrastructure, experience, skills and knowledge. Moreover, legitimate research directives may give priority to promising areas, such as nonlinear dynamical systems theory, quantum security, quantum electronics, photonics, nanophotonics, fiber lasers, solar power, terahertz waves and wireless communication systems. A smart future society may afford broadband internet, navigation technologies, sophisticated 3D maps, radar, sonar, satellite communication, autonomous vehicles.

Future smart conceptions may imply smart grids and microgrids, smart consumption, glass fibre networks, smart convenient home appliances, distribution management systems, synchronised grids monitoring, pioneering progress in infrared imaging etc. (El-hawary, 2014). The economic value of the ICT will also become very significant, because of the need to estimate and control the costs and the gains of alternative ways of production, fuel economy and reduction of emissions, of testing and introducing modern applications, and inventing new ones, as well. However, the fact remains that even today the part of ICT in the total emissions is small.

The freedom to initiate and participate is critical in smart societies. The building of a brighter future is a collective effort that refers to communities, citizens, local leaders, depending on their need for community change, at the right time for investment, with cooperation, focusing on community strengths, practicing democracy, preserving the past and inventing a smart future (Morse, 2004).

Chapter 7

Scientific conceptions of the universe

Within the realm of daily life, people don't reflect profoundly either to the large scale of the universe, i.e. to the machinery that generates sunlight, or to the very small scale. In ancient cosmology, the arguments in favour of a round spherical earth were based on the evidence of eclipses caused by earth's position between the sun and the moon. Further evidence was provided by the earth's shadow on the moon, which is always spherical, as Aristotle knew. In addition, the Greeks had realized that the North Star, lying over the North Pole, appears lower in the sky when viewed in the south than it does in more northerly regions. Moreover, the belief on the spherical shape of the earth was being corroborated by the fact that the sail of a ship is visible before the hull (Hawking and Ellis, 1973).

In antiquity also, the atomists developed the theory for the infinitude of space, for instance, Lucretius "conceived space as endowed with an objectively distinguished direction, the vertical," as Jammer (1993: p. 13) notices. However, the false part of Aristotle's and Ptolemy's theories was isolated only in 1609, when the observation of Jupiter's satellites proved that everything did not have to orbit around the earth. After that discovery, crystal celestial spheres and natural boundaries to the universe lost their traditional meaning.

Nonetheless, the most interesting part of the new theories was the belief that the planets were made to orbit the sun, driven by magnetic forces. This opinion was transformed by Newton into the law of universal gravitation, which resulted also to the supposition that there is an infinite number of stars. Additionally, Galileo's experiments and Newton's first law expressed the revolutionary conception that the natural state of a body is motion, rather than rest; while the real effect of a force is always to change the speed of a body, rather than just to set it moving.

Obviously, this theoretical change caused certain difficulties: Although Newton's law of gravity says that the attraction of a star is exactly one quarter that of a similar star at half the distance, the lack of an absolute standard of rest meant that an absolute position in space is inconceivable; thus, absolute space, is relative.

Furthermore, in 1676, the Danish astronomer Ole Christensen Rømer discovered that the light travels at a finite, but very high, speed (*Démonstration touchant le mouvement de la lumière*). Jupiter's moons offered observable differences to support the finiteness of the speed of light.

Momentum and inertia. The overcoming of Aristotelianism

Medieval scientists like Buridan and Oresme assumed that the earth rotates around itself, but yet there was no question of removing the earth from the centre of their worldview. Buridan and Oresme kept simply in mind the daily rotation of the earth around its axis, in order to avoid the need to consider the entire celestial sphere as rotating. Specifically, Oresme considered for reasons of economy, that it would make more sense if the earth rotates. However, there was no physical evidence of rotation, because the observed motion is only relative and not absolute. Eventually, Oresme retreated to the opinion that the earth is immovable.

The concept that aided science to unshackle from Aristotelianism and to discover the physical principle of inertia was the *momentum*, initially introduced by Oresme and Buridan. The latter used the new term *impetus* to denote the applied force. He tried to distinguish the momentum from the movement caused. Buridan argued that the amount of momentum can be measured by the speed times the quantity of matter of the moving body. He explained the acceleration of a falling body with the assumption that, as the body falls, its gravity constantly generates additional momentum to the body and, as the momentum increases, causes an increase in speed.

The first theory of that kind was proposed by Philoponus and Avempace, namely that speed is force minus resistance. The second theory stems from a sentence in a passage of Averroes: the speed is equal to the ratio of force minus resistance to resistance. The third theory represents the traditional interpretation of Aristotle's view: the speed is equal to the ratio of force to resistance. The one who undertook to resolve the conflicting views was Bradwardine. He refuted all three aforementioned theories, noting their absurd and unacceptable consequences. The first was rejected, because it is contrary to the finding of Aristotle, that the simultaneous doubling of force and resistance leaves the speed unchanged.

The third theory fails because it does not provide a zero speed when the resistance is equal to the force or bigger than this. Bradwardine proposed an alternative 'law of dynamics:' He considered that the speed increases arithmetically as the ratio *force to resistance* increases geometrically. When speed is doubled, the ratio force to resistance should be squared. In order to triple the speed, the ration force to resistance should be cubed and so on (Lindberg, 2007).

The attention has been often called to the quantitative compatibility between the impulse (impetus) of Buridan (speed times quantity of matter) and the modern concept of momentum (speed times mass). Nevertheless, while the momentum of Buridan is identical with the cause of the conservation of motion of the projectile, the modern notion of Momentum is a measure of a movement which does not require a reason for its conservancy, as far as it does not meet resistance.

Scientific revolution developed through critic both to Aristotelianism (Copernicus, Brahe, and Kepler critically embraced the influences they obtained in universities) and to Cartesianism. The experimentalist Huygens carried on the dialogue with Van Schooten, his professor in Leyden, upon Descartes' worldview, and supported a mechanistic corpuscular philosophy, giving much more emphasis on physical observation than Descartes. The controversy about Cartesianism was a *springboard* to progress.

The most superb innovation of the early modern science was the emergence of mechanistic physics, through a new system of natural philosophy that incorporated an advanced mathematical theory. That became gradually true. Leibniz agreed with Descartes that motion in an infinite universe, with no vacuum, implies, first and foremost, an infinite number of vortices (Gaukroger, 2002), an idea conceived firstly by Leucippus. The space was filled with an ether of ultrafine particles and the rotation of the Sun caused cyclical movements, vortices, in the ether, which pushed the planets around the sun as the boats moved by a whirlpool. The problem of gravity is associated with the theory of vortices supported by Descartes, Leibniz, and Huygens, and with the difficulty to accept an action from a distance, because it is not observable.

Leibniz considered as cause of both gravity and planetary attraction the cycloidal motion of the ether,[9] a very thin fluid, from traction spokes, which disrupts matter with countless ways, from all sides, with the result, however, that the motion of the planetary bodies converges at a certain level and in a particular area, and the more massive bodies move toward the centre of the vortex.

In their correspondence, Huygens and Leibniz, while discussing the theory of Newton, they made clear their differences with him. The planets do not just move in ellipses, said Leibniz, but they also move altogether at the same plane, in proportional directions around the sun. Therefore he rejected the Newtonian attraction since it would fail to produce movements in a wider and not limited region of the three-dimensional space. At another instance, Leibniz wrote prescriptively to Abbe Conti that the most different causes engage with each other, in our explanation of gravity, namely the spherical radiation, the magnetic attraction, the release of spinning material, the inner motion of fluid, the circulation of the atmosphere, all together conjoin in the production of centrifugal and centripetal force.

In *Tentamen de Motuum Celestium Causis*, Leibniz, leaning on Kepler's laws, described the liquid spheres that move the planets. However, Leibniz understood the inertia as resistance, a force which is opposed to motion. Another problem of modern physics was the concept of force itself. According to the mechanistic understanding of the concept, force is the pressure or thrust of a body to another.

Descartes, Gassendi or Boyle would not disagree with this wording. Descartes, who tended to consider the moving particle as a cause, reported on the 'force of the motion of a body.' He insisted that the force with which a body acts upon another, or resists the effects of it, consists in this alone, that everything does everything it can to remain in the same condition in which it is. Descartes was identifying force with inertia. The principle of inertia was one of the cornerstones of the mechanic philosophy of Descartes. All natural

---

[9] De Causa Gravitatis, et Defensio Sententiae Autoris de veris Naturae Legibus contra Cartesianos.

phenomena were caused by moving material particles which are inert by definition. No cause is needed to *continue* the motion of matter. The motion can be transferred from the one body to another upon impact but remains indestructible.

Interactions between bodies are caused only by impact. In this context, it was easier to challenge the view of Galileo that the inertial motion is a circular motion around a centre of attraction. Descartes concluded that every moving body always tends to follow rectilinear course. It moves in curve, only if another body diverts it from its course (Westfall, 2004).

Descartes and Gassendi contributed to the debates about the concept of inertia and were the first to suggest that the inertial motion should be rectilinear, and that the circular and curvilinear motion of the body are caused by external factors, meaning that those bodies that move with circular motion are found within vortices which create two movements: the centripetal, which results to the effect of gravity, and the centrifugal, which results in the effect of light. Therefore, Descartes was explaining the motions of the planets with the theory of vortices.

Newton proved, however, that vortices can never imply a planetary system that moves with Kepler's laws. He formulated thus the principle of inertia as follows: Everything remains due to his own nature in the situation in which it is found unless it would be changed by some external cause, therefore a body having been moved, maintains for ever the same speed, the same amount and direction of his motion. The same applies to rotating bodies, as the principle of conservation of angular momentum outlines.

The Newtonian Mathematisation explained the absolute circular motion and the absolute acceleration during rotation with the second law for the conservation of momentum and of angular momentum, whereby the *momentum*, the *angular momentum* of a material point, or the *primary torque* of an inertial system varies only under the influence of external forces. On the contrary, internal forces can only change the angular momentum of sections of the system and the angular velocity. The main inertial torque in the centre of the mass is always the smaller.

We can logically infer that Newton stressed the issue of *absolute acceleration* during rotation, because a simple variation of the angular speed, namely a relative speed, may, in accordance with the second law, be the result of internal forces. Nevertheless, absolute acceleration increases the moment of inertia, the angular momentum of the system overall. The difference from Leibniz is striking.

Mathematisation allows for a simple explanation of the entire global movement at the base of Newtonian laws and the principle of inertia. The Newtonian principle of inertia states that it is not necessary to have a mover in order to produce movement. That is absolutely natural to continue a linear or circular motion endlessly if the moving body does not find obstacles in its path. The law of universal gravitation states that there is a gravitational force that all bodies dispose, depending on the various quantities of matter they contain. The body tends to retain the state of motion and this is called inertia, due to the conservation of momentum. Momentum is equal to the speed times the mass.

Philosophy and Scientific Revolution

Floris Cohen (1994) pointed to the important contribution of Alexandre Koyré (1939; 1957; 1968) in underscoring the consequences of the significant influence and "integration" of Neo-Platonism in the scientific revolution. More specifically by undertaking the defence of Plato in the same context. Even in the educated classical Athens, the apparent daily motion of the sun from east to the west it was difficult or impossible to be regarded as a mere appearance, misleading yet. At least with the means of those times. Plato's insistence on the dismantling of phenomena uncovered the planetary movements also as a problem. By insisting on the cyclical nature of the movement, Plato highlighted the value and the need of mathematical conceptualisation. This approach focuses on the priority of knowledge over presumption (δόξα), namely the phenomenon. Therefore, Plato directed a challenge to the mathematicians, to examine this problem methodically. Eudoxus accepted this calling of his master. It is no coincidence that the alleged mathematical error of Aristarchus, to consider the center and the surface of the sphere as proportional sizes, had undermined the cogency of his thesis, as it was suggested by Archimedes in *The Sand Reckoner.*

During the seventeenth century, the mathematical ideal inspired the European scientists and the mathematical research became decisively purposeful, with the harvest and exploitation of the observations made with the telescope, the microscope, the thermometer, the barometer, the air-pump, and the pendulum clock etc.

The mathematicians were fascinated by the fact that Plato presented his views in dialectic form, inventing thus the ideal form of the development of the scientific community. The dialectic in the form of *dialogue* according to Socrates, the conceptual division according to Plato (Sophist) and the question according to Aristotle, motivated the establishment of collective scientific projects inspired by masterpieces such as those stemming from the interrelated efforts of Euclid, Apollonius, Archimedes and Heron. The form of the dialogue was selected also by Galileo, who even used the Italian language for his purposes.

From the aspect of content, the dialectics was fruitful to the comparatively mingling Neoplatonist environment of the Renaissance: either, as in Aristotle, dialectics is contrasted with proof, because dialectics regard chance, or, as in Plato, it is identified with the process of dividing concepts for making definitions. A closer inspection, of course, from Kepler to Galileo and from Huygens to Newton, enables us to discern the process of the refined assimilation of Neoplatonic elements, the emergence and overcoming of the mechanical worldview, and a more independent epistemological placement of the scientists of the seventeenth and eighteenth century.

The readiness with which Newton (1968; 1729) distinguishes the mathematical figure of ellipsis from the corresponding equation, shows that in the intervening years the mathematical ideal came closer to reality. The dialogue, even the daily correspondence, refers to abstract formulas, mathematical equations. Once more, Platonism is being surmounted, as the equation and its variables are bearing largely the elements of observation. Therefore, we identify the first discontinuity. The same applies for the influences by Democritus, which are enriched with the concept of force but are being transgressed by Mathematization.

In general, we cannot overlook the functional but discontinuous and no permanent influence exerted by different aesthetic, cultural, religious or philosophical currents, in their connection with the historical needs that fuelled the scientific discoveries. Through ruptures and affinities, confronting new demands each time, similar mingling currents, such as Neoplatonism, are being articulated, impact and fade away, after they have incorporated and transformed the more radical Platonic, Stoic and Aristotelian elements.

The Neoplatonic philosophy, enlightened with the concepts of invention, dialectic, definition and Mathematization, was a specific and favourable historical effort, an ideological component with valuable theoretical implications, which contributed to the liberation from dogmatism, religious or philosophical. On the other hand, different character acquire social movements that can continuously affect in an uninterrupted and organic manner, not only functionally, the science, directed by more immediate intentions.

According to Crombie (1992), a scholar dedicated to the internal historical development of science, the improvement of experimental and mathematical methods during the thirteenth and the fourteenth century caused a movement which became so pronounced until the seventeenth century, that it was called Scientific Revolution. Crombie argued that during the late Middle-Ages an appropriate methodology was being processed, i.e. the experimental methodology of modern science.

Other scholars instead, such as Francis Bacon, argued that the period between antiquity and early modernity was an era of misfortune for science, while Arabs and Scholastics shattered science. For this reason, scientific revolution may be distinguished from Middle Age, because by seventeenth century, according to what Francis Bacon states in the Novum Organum, we have: a) liberation from the largely inapplicable mental experiments, b) systematic experimentation, and c) careful observation, which was favoured by important innovations such as the telescope and the microscope. Many others agreed with Bacon, such as Voltaire and Condorcet.

Anyway, the relationship between faith, science, philosophy, and scholasticism has been a fascinating theme in philosophy of science (Duhem, 1954; Hooykaas, 1972; 1999). In practical terms, Duhem and Dijksterhuis were, of course, correct when underlining the importance of the introduction of the concept of impetus (momentum). Here the decisive factor was not only the latent scientific accuracy but also the mediation of the definition, a comparative philosophical contribution, which indicates that scientific reason requires simultaneous work on both empirical and theoretical components. Therefore, not every kind of anti-scholasticism is progressive, let alone anti-Aristotelianism itself. As well as any philosophical stream itself, regardless of the social environment.

Copernicus and Kepler

Sometimes the work of the historian of science is insurmountably difficult. How could we deal with cases like the incidents of the edition of *De Revolutionibus orbium coelestium* and the censorship exercised by the Lutheran theologian and friend of the author, Andreas Osiander? The book described a mathematical and physical model that presupposed the revolution of the earth and the planets around the sun and the rotation of the earth on its axis. After the departure of Rheticus from Nuremberg, the manuscript fell into the hands of Osiander, who decided, not well-intentioned, to alter it in various ways, without informing and requesting approval from Copernicus himself. Osiander is supposed to have put the word "Hypothesis" on the title page, deleted important passages and added his own propositions. For instance, the foreword, with the title: "To the Reader; Concerning the Hypotheses of this Work," is assigned to Osiander. In such occasions, it became necessary to organise simultaneous literal and historical scientific research in order to ascertain which exactly the interventions were.

Similarly, inextricably confusing problems occur with the study of the works of antiquity. The complete recovery of the original ancient texts is sometimes impossible. Herculaneum is the only ancient library that its books are today extant, while the Arabs acquired a quantity of original Greek manuscripts, which they translated and commented. Thus, through the Arabs, works of Heron, Stephanos of Alexandria, Hippocrates and Galen, many works of Aristotle and others were transferred to Europe. However, the contact of the Europeans with the Greek texts was not continuous and many translations are incomplete. Our main sources for ancient and Arabic science was Adelard of Bath, Gerard of Cremona, William of Meerbeke etc. In the period of the scientific revolution, the scientific texts of the antiquity had valuable influence. An effect that has not yet been investigated in depth. For example, the study of Apollonius' works helped Kepler in the formulation of the laws of planetary motion. Kepler, however, was taking his careful distances from antiquity: He rejected the ancient insistence on circularity and tended to accept elliptical orbits, which were conformed to his observations and his preference for regular solids. With the adoption of the eccentric circle, he liberated science from the system of the epicycles.

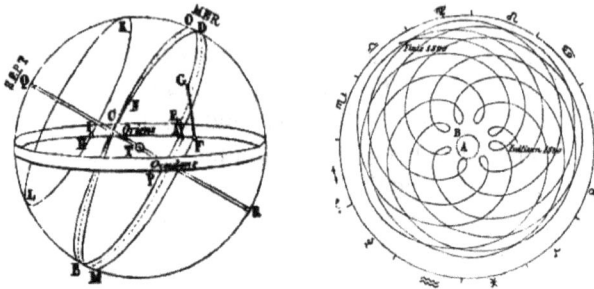

7a The orbit of Mars was a main theme of disputes between supporters of the Ptolemaic, the Tychonian and the Copernican system (Johannes Kepler, *De Motibus Stellae Martis*).

After eight years of observations and heuristic estimations of the orbit of Mars, Kepler concluded that the radius of curvature was measurable within a sinusoidal function. This simple mathematical element, describing a uniform physical variation was definitively outweighing metaphysical cosmology! Quite simply Kepler managed to make recordable the orbits of the planets with a simple conical curve. The simplicity of the ellipsis was demolishing the system of the epicycles.

The *Conics* offered an uppermost contribution to this achievement. They belong -along with the book of the astronomer Menelaus on curved triangles- to the projects of antiquity that enabled the overcoming of the Ptolemaic system. The complete work of Menelaus and much of the *Conics* are extant only in Arabic. Again there are more questions and only a few certainties. How those texts became available to the astronomers of the seventeenth century? To what extent they were aware of them, they taught and discussed their content?

Kepler confessed the influence of the *Harmonics* of Ptolemy, with respect to the regular polyhedra. The manuscript of that work arrived in his hands, according to Kepler, through the Chancellor of Bavaria. The idea expressed in *Harmonics*, the heavenly harmony - that the world is mathematically symmetric and that the celestial movements are harmonic- was advising for the discovery of the laws of planetary motion.

Under those influences, a philosophical Kepler was shaped, who touched the issue of universal harmony, where the keys of the musical scale, for example, the tons of the musical system and the types, major and minor, of symphonies, are expressed, according to Kepler, with respective planetary movements. Copernicus dedicated general references to Ancient Greeks and the Arabs, like Al-Battani, as scientific precursors in grasping the theorems for the curved triangles that we find in *De Revolutionibus*, from the 14th chapter of the first book and further. He calls them 'spherical triangles' and, with great accuracy, he uses them in the calculation of the parallaxes of the sun and the moon. He doesn't attribute their origin to Menelaus, but he refers to proofs of Euclid. Copernicus mentions Menelaus on another occasion, in the second chapter of the third book. Comparing now these so contiguous types and the influences they had obtained, we easily distinguish two distinct types. The educated intellectual of the church, who is constrained by the requirements of the ecclesiastical regulation. Without escaping surveillance, he suffers as author and scientist. On the other hand the scientist, as a common man of Central Europe, having an extremely complicated worldview, metaphysical admittedly, but keeping in centre the Pythagorean faith to the mathematical harmony. Born in an era of a titanic spiritual and social conflict, with astronomical evidence at hand, he needed to express firmly his beliefs in the most crystal clear way.

Copernicus' Dedication to the Pope and Galileo's Letter to Christine

In their efforts to associate with the powerful ones and to explain their theories, while coping with canonical and philosophical conservatism Copernicus and Galileo used a series of modern but mainly conservative arguments. Their dialectics were less contentious and more rational and reformative, especially that of Copernicus. Similarly, we can distinguish a lot of differences that decisively outline their diverging fortunes. Copernicus, in his Dedication to the Pope, with some introductory expressions, displays a behavior almost beyond modern belief, since today we are not addicted to a limited access to knowledge. Many scientists, according to the cleric Copernicus, were used to revealing the secrets of their philosophy only to relatives and friends, not written but orally. It seems that the Polish astronomer, at least wavers in front of the technical capabilities of the invention of the printing press, desperate which path to choose. He appears self-contradictory when he confesses that because he was afraid, he did not earlier announce his views. Copernicus uncovers clearly the origin of his despair, when he prefers, even at the last moment before his death, to avoid to defend his work openly: In the end, he allowed friends to undertake the adoption of the project, something they begged warmly for a long time. The explanation of an *alienation of the scientist from his work* might be appropriate in such circumstances, along with the facts of Osiander's intervention in the writings of Copernicus (1992).

This kind of conservative arguments and attitudes are manifestations of the timid steps to the mature stages of science, bearing however heavily upon them the signs of social and religious oppression. Copernicus, as we see, was forced to retreat to traditional and almost backward models of contemplation and representation of the social environment, in order to justify the overly cautious emergence of his work. Comparable opinions, degrading for the perceptions of the "ignorant" majority, held Galileo, as well.

Nevertheless, the point of rupture is quite carefully unveiled, when Copernicus retrieves from oblivion the names and theories of Hicetas, Ecphantus, Philolaus, and Heraclides. The ancient tradition of heliocentric theories is treated with a certain pragmatism and the author is looking for authors that will assist him, in the present, to draw a middle ground between innovation and tradition. The ancient texts might legitimize modernity. However, he omits to present Aristarchus, at the moment, the scientist who was convicted of insulting the gods!

Copernicus confesses often his debt to ancient science in moral and scientific ways, such as in the tenth chapter of the first book, where he observes that Euclid proved in his Optics that between two objects moving with equal speed, the more remote seems to go slower. In the field of measurement and observation, the requirements of continuity, which shaped centuries of systematically collected data, recall again, for internal purposes, the invocation of tradition. Obviously, the eccentric and the epicycles have introduced many ideas which contradict to the basic principles of uniform motion (Copernicus, 1992).

However, the originality of the author becomes manifest, when he confesses that he was occupied with the confusion in the astronomical tradition about the appropriate way to deduce the movements of the spheres of the universe. It was undeniable that the complexity of the models of astronomy was not compatible with

the simplicity of uniform motion. Copernicus, however, was interested only in the state-of-the-art information and reported on innovation with an intelligible way. This is manifested by his intention to put his work available to scholars of astronomy; the only
reason that prompted him to look for a different system, from
which the motion of the spheres of the universe is deductible, is the
finding that astronomers do not agree between them in the investigation of that object. However, we cannot know whether the con-
ciliatory, rational and organizational manner of his intervention
and his contribution to the scientific community, would be possible
at all, without proper ecclesiastical support, in his effort to solve the
problem of reforming the church calendar. In addition, the authority of the Bible is not an object of dialogue, according to Copernicus, and the question of astronomy is completely independent of
faith. This, as we will see, is a significant difference from Galileo.
Unlike the Bible, the antiquity is not an authority; the recursion to
scientists of the antiquity is carried out because it renders a field of
fertile theories, concerns, and examples.

According to Galileo, Aristotelianism is a hindrance, while the authority of the Bible is treated in a rather heretical way. Let us re-
member that heresy connotes choice and hereticalness means
eclecticism. Galileo seems indeed consciously heretic when he
opens to Christina the conversation on the interpretations of the
Bible. This "impropriety" equals to an unprecedented for a mathematician intervention in theological thinking. He does not hesitate
to entangle himself in theological issues, by stating that if one was
always limited to the literal meaning of the interpretation of the
Bible, one might fall into error. He considers useful for his purposes
St. Augustine's advice. Nevertheless, Galileo gives rise to a serious
conflict, by expressing a risky opinion upon Bible hermeneutics in
this letter, which ends up with open question marks. As pointed out
by Crombie (1992), Galileo felt that mathematical science is a
method to read the real book of nature. One wonders yet, what
symmetric and comparable aspects may faith and science have
when belief is contrasted with faith, though belief includes the
probabilities, against which faith is completely inhospitable? The
faithful are not conquerable by rational or empirical counter-
arguments and they may not admit taking account of probabilities.
Their principal, favorable argument is faith. The arguments used by
Galileo are conservative and apparently follow the line of Aquinas,
to draw up a clear demarcation between faith and reason. The opponents of Galileo, as the Italian mathematician contends, made

the fatal mistake to dress their arguments with passages from the Bible, which were not properly understood and which were totally unsuitable for their purposes.

A logical objection is that the limits between science and religion were not in that historical moment correctly recognized. Nor the way in which Galileo acted was effective. In addition to this letter, which appears detached from the whole of his work, Galileo chose elsewhere, in Dialogues, another line of argument, anti-Aristotelian, without elaborating on faith. It is paradoxical that he recollects his reports upon faith in the letter to Christina. I tend to think that he selected this dangerous field, neither motivated by his interest in theology, nor wanting to recall the strict distinction between reason and faith, but because he sought how to confront the editions of those books that supported misleading views, such that the moon is light-emitting.

He hoped to prove that he was acting with great reverence, insisted Galileo. He supported Copernicus, who should not be convicted without being read. The astronomer of Pisa appeared, notwithstanding his familiar scientific behaviour, to invoke higher authorities. Galileo even declared that he wanted to help the Holy Church to take a decision on the Copernican system, in order to adopt it and let use it as the officials may think best. That's because, once again, the authority of the saints and the educated theologians assures us that the truth of the Bible cannot contradict to the strict sense and experience which are constituting human knowledge. As pointed out by Galileo the opposite attitude would ultimately be hurtful for the church. We step back into the same stressful thoughts about the alienation of scientist from his work: Galileo warns his enemies to keep away from tearing and burning his book, while he never sought nor wanted to derive from it a benefit that is not consistent with the pious customs and the Catholic Church. However, he notes that what must be really avoided is to pervert the biblical passages which relate to matters of faith, or morality, or to the teaching of the Christian doctrine. That was the decision of the Council of Trent at its fourth session (Galileo, 1953).

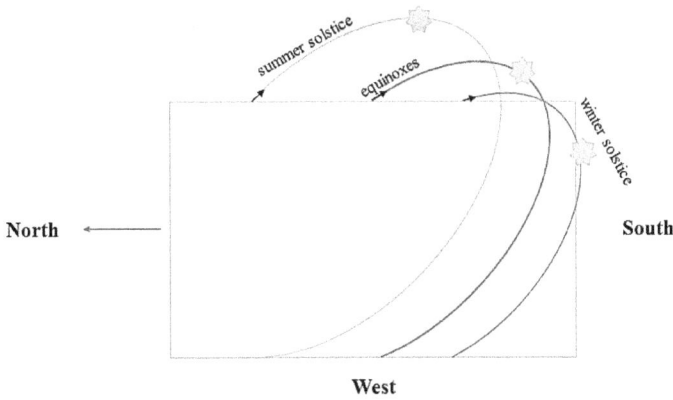

7b The paths of the sun on the solstices and equinoxes (Author's design).

The motion or immobility of the Earth and the Sun are neither an issue of faith nor a breach of ethics. Moreover, there are many open interpretations of Joshua's miracle but, above all, the miracle would imply, according to the Ptolemaic system, either the more rapid movement of the sun, or the immobilization of the whole universe!

The ancient tradition offers evidence of the continuity and legitimacy of Galileo's views when referring to the controversial, for theologians and philosophers such as Augustine, hypothesis of heavenly motion. One of the empirical problems is associated with Sun's immobility as opposed to the motion of Earth, a view supported by Pythagoras and all his followers: Heraclides of Pontus, Philolaus, the teacher of Plato, and, according to Aristotle, Plato[10] himself. This is, I believe, the most important of the traditional influences that Galileo incorporates in his worldview, but at the same time his most active modernist standpoint, in this letter to Christine. This view was shared, according to Galileo, by Archimedes, Aristarchus of Samos, Seleucus the mathematician, the

---

[10] Indeed, Plato in *Timaeus* while accepting that the earth is located in the centre of the shaft of the world, he considers the universe as rotating.

philosopher Niketas (according to Cicero's testimony) and many others like Seneca. Galileo probably revoked the model with the vortices, which was supported by atomist philosophers. The image of the earth in the center of a rotating world is similar to the image given by the filings of iron floating and rotating in the centre of a vortex, in Huygens' experiments with the rotating bucket. This delivers a familiar natural phenomenon, which fits as it is clear with a later Cartesian worldview, inspired also by the image of a Platonic earth.

There is a whole range of naturalistic and cutting-edge arguments in Galileo's letter. Galileo starts touching his own discoveries and those of Copernicus, such as the phases of Venus, and writes that the increase in the number of knowledge is an incentive for research, consolidation, and development of the arts, not for their shrinking or their destruction. He recognizes the validity of the geokinetic views and considers as favourable the occasion for their adaptation. Of course, Galileo notices that these views have stirred up against him a great number of professors, but he recalls the Herculean vigour and the admirable genius of Copernicus, in the appropriate circumstances of the correction of the ecclesiastical calendar, sponsored by the pope of his times.

There are notable examples of natural philosophy in this letter. For Galileo, Nature is relentless and unchangeable; it never violates the laws that have been put to it, nor cares at all if the obscured causes and ways by which it works are being understood by people. Another modernist argument is the scientific consensus gained by the books of Copernicus and his own, according to eminent mathematicians. Also significant is the zeal with which their works are being read and comments are being collected on the ratification of their positions.

It seems however that the letter's main objective was to convince Christina that there is a theological interest in the distinction between reason and faith. Galileo supports that God wrote not one but two great books, the book of Nature and the Scriptures. Galileo's arguments, however, did not yield the corresponding substantiation in terms of reason and science, but still, he remained attached, emotionally perhaps, to the religious field, eventually failing in his objective to defend the geokinetic system and avoid the ban of his books (Galileo, 1953).

An excerpt from the Dialogue Concerning Two Sciences

The conversation on the First Day (*Dialogues*, 50-51) provides a documentation of the view that, from sixteenth century, technology acquired a more significant position in the scientific work, since scientists began to deal systematically with technical problems offered for analysis and research, to experiment with technological equipment and its use as scientific instruments, and with new observation areas. The technological environment in which Galileo took action facilitated the excellent scientific understanding of technical developments which imparted to his work.

Salviati (*Dialogues*, 50) praises the constant activity of the Venetians in a wide research field, especially in engineering and shipbuilding. All types of instruments and machines are manufactured with expert programming, based on experience and observation by sufficiently skilled and clever craftsmen. Sagredo agrees with Salviati and distinguishes the influence of masters, with their incomparable works, including careful experimentation, such as with the overload of large ships.

Salviati points out: a) the comparative ease of manufacturing larger machines such as with the case of clocks that are more punctual on a large scale due to the diversification and the imperfections of the material, and b) the greater durability of small engines, because of the resistivity and the reduced firmness of large machines. These observations make Sagredo understand that geometry is not sufficient for the construction of the same machine from the same material in different sizes.

Then, they make similar observations for the strength of rods, trees, beams, columns, leading Sagredo to the formulation of the mathematical problem which will occupy them [54]: they question why the strength and the resistance are not multiplied in the same proportion with the material; why, as has been observed in some cases, the strength and the resistance to breaking increase in a greater proportion than the amount of the material? In parallel, they consider the technical problem of the nature of the material responsible for the cohesion of solid bodies and they observe, for example, that the ropes are cut because their curled fibres glide rather than because the fibres break.

In this way, it is, finally, outlined what Galileo called a new science, which is based both on engineering and material, and on mathematics. This new science is based on solidly founded observations of daily technique, sometimes as simple as that of the boating knot, the spring, and the durable packing rope. Similar techniques were used in the machines of the winch, in lifting an anchor from the sea bed and in launching loads into mines.

A factor of the strength of materials, according to Salviati [59], is the Aristotelian aversion to vacuum, the horror of the void that is a form of resistance of the material to the emergence of vacuum, which has as a consequence the sticky nature or the viscosity[11] of certain natural bodies. They consider, for this reason, dyed and extremely smooth plates made of metal and glass that slide one over the other, because there is no resistance, as the sticky material is missing. Sagredo adds that actually, the movement in a vacuum is not instantaneous, as the one plate tends to follow the other in motion, and that this (thought) experiment shows that only for a moment vacuum is created between the plates. The vacuum is created sometimes by violent movement. Sagredo, however, is not satisfied with the explanation of the resistance to vacuum, because, as he observes, the cause precedes the effect.

This gives the opportunity to Simplicio to intervene, in order to support the Aristotelian view on the issue. Sagredo yet insists that the aversion to vacuum is no sufficient reason for the emergence of any kind of resistance. His argument, in its second part, is that it would be wrong to isolate the aversion to vacuum as the overriding cause, between a series of causes. However, Salviati supports the opinion of Aristotle and explains the resistance of the water to scattering, just with the *horror vacui*. He refers, in support of his claims, to the works of the Academician, implying Galileo.

The experiment carried out by Salviati [62-63] (with two unequal cylinders, fitted one inside the other, while between them water is

---

[11] The same concept was used by Leibniz to interpret the inertia as resistance, absolute on regard of the viscosity of the fluid and relative on regard of its density (Bertoloni Meli, 1993). In their correspondence, Clarke insisted that the different resistances of mercury and water are caused by their different densities, and there is necessarily therefore more vacuum where resistance is less. Leibniz found an opportunity to "correct"... it is not so much the quantity of matter, as the difficulty to find space [viscosity], which creates resistance.

injected and air is removed carefully; when the cylinders are being inverted, they remain contiguous, unless we mount weight to the small, in order to separate them) is indeed impressive and makes Simplicio concerned, since it includes the skilful removal of air. The experiment embarrasses Sagredo also, who is forced to retreat. The dialogue continues comparing the winch and the pump, whose function observed Sagredo, in a tank. The most important issue emerging from this debate is the declaration of atomism, where Salviati takes a clear stance against Aristotelianism, accepting the hypothesis of the existence of infinite small vacua. These micro-scopic vacua exert a force that is in the limit so great, as to cause phenomena such as the melting of metal from the heat. Nevertheless, this is an inaccessible to experience explanation. Therefore, Galileo returns quickly to mechanics, to Aristotle and to sizes measurable by instruments of that time, and he examines the problem of the calculation of the movements of cycles and hexagons with the same radius.

The instruments and the machines that Galileo analyses in the commented excerpt come from all areas of production and techno-logical application, from mines, workshops, shipyards etc. The measurement and the Mathematization are implemented precisely in the frames of the means of production. Even in the case of as-tronomy, the introduction and improvement of the magnifying lens improved also the measuring instruments, and simultaneously the use of the telescope was extended from the sea to many other uses, serving many practical purposes and facilitating the feedback of the astronomical findings (Galileo, 1957; 1968).

The study of the historiography about Galileo offers further documentation of that particular sense of a *new science* that keeps pace with technology. It appears as a replacement of an idea about the nature by an alternative idea; of one worldview by another. It has been described also as the *technically produced nature*, by Flo-ris Cohen (1994). Rationalistic perspectives, such as that of Alexan-dre Koyré (1957), emphasized on the mathematical abstraction and used as an example the invention of the pendulum clock by Huy-gens. Christian scientists, such as Reijer Hooykaas (1972; 1999), stressed the undermining of scientific authority and the strengthen-ing of the empirical, naturalistic historical methodology. Hooykaas (1983) relied on the original study of the reconnaissance travels of the Portuguese. The approach of Ben-David (1971), while recogniz-ing the impetus given to science by the widening of the horizons of experience, insisted that seventeenth century was a rare exception in the slow, irregular and patient advance of science.

Mathematization and Scientific Revolution

Francis Bacon proposed many useful ideas for the development of scientific activity, such as the distinction between: a) the illegitimate attempt that hovers from the senses and the particulars to the more general postulations, and b) the proper methodology that generates the axioms from the senses and the particulars, rising with a gradual and unbroken intensity, for the purpose of arriving at the most general axioms, last of all. Bacon (1989) also was the first scientist who articulated, according to what we know, the concept of the *crucial experiment*. His omission or failure was that he did not give an equal consideration to Mathematization. Therefore, the potential completeness of observations and experiment, the gradual and coherent progress remains undeveloped, an ideal without user manual, one would say.

On the contrary, both Galileo and Kepler seek to satisfy the Cartesian requirement for a mathematized science which is based solidly on precise and clear ideas, derived from experience and logic. The liberation of logical thinking and the development of scientific activity, after the Renaissance, became feasible and were specifically realised, only through the compliance of science to mathematical standards. This is, I believe, in accordance with the rise of the broader social demand for consistency between beliefs and experience, between theory and observation. It presupposes also the requirement for reasonably demonstrated scientific thinking, while with this process all irrational elements are being criticised. Both technology and economic thought made this development possible. We could even consider the development of the arts of measurement, digitization, and calculation as an indication or measure of rational and scientific progress.

The new natural science may be divided into four stages: a) identification and careful investigation of phenomena, leading to the recognition of their characteristics and the measuring of their proportions, with the recommendation of hypotheses by analogy based on *discontinuous* or *incomplete* series of empirical data sets, b) gradual organisation of deductive procedures based on *finite* or *continuous* series, for which a mathematical formula is available, and c) inductive procedures, which refer to *infinite* series of data that we have satisfactory identified and mainly approach by statisti-

cal methods, d) *experimental test* through which the above processes will be verified or they will be falsified. Galileo incorporated in
his work various forms of experimental measurements (In the second part of the *Dialogues*, for example, where dealing with the consequences of the mathematical concept of uniform accelerated
motion and the proof that the fall in a vacuum is uniformly accelerated). However, the inability to experiment in a vacuum forced him
to confine himself to the examination of the effects of uniformly
accelerated motion, as they appear in inclined planes.

Galileo's methodology was often hypothetic-productive and not
demonstrative, as he would prefer. He even found out with experiments -which he repeated hundreds of times- that in all inclined
planes the displacements are always proportional to the squares of
the time of fall. Remarkable is the measurement of the distances
gone through, in equal intervals, by a body falling from a stationary
position, which distances are found the one to another in the same
ratio as the odd numbers beginning with unity [Dialogues, 190]. As
pointed out by McMullin (1998), the method of Galileo is sometimes demonstrative and other times, when he has no access to the
diligent observation, hypothetic-productive. An alternative description of Galileo's methodology may be *mathematical realism*, which
requires adequate competency in geometry, as Biagioli (2003) proposed.

Kepler also developed a mathematical science, which is in pursuit
of only true premises and not merely of hypotheses, claiming thus
certain knowledge. The *Mysterium Cosmographicum* (1596) was the
first work that supported the Copernican theory after the *De Revolutionibus*. Once more, in 1618, Kepler informed his readers that
they should know that the ancient astronomical hypotheses of
Ptolemy are to be neglected entirely from their discussions and
minds because they are not verified by the physical bodies and their
movements.

In his *Harmonies* [5, 3] Kepler insisted with absolute certainty that
the theory of the heliocentric universe is true, and asserted with
certainty that the planetary motions are elliptical with the sun at
one focus. Kepler nevertheless considered, as in the *Mysterium
Cosmographicum*, twenty-two years earlier, that the number of
planets other than Earth, and the number of orbits around the sun
was five, as many as the *regular solids*, as they were described by
Plato and Euclid. The planetary movements, according to Kepler,
follow specific harmonies, as Pythagoras realised.

In his early work on Mars, Kepler had shown that equal daily arcs in the same eccentric circle are not traversed in equal speed, whereas the differentiating delays in equal parts of the eccentric, maintain the ratio of their distances from the sun, which is the source of motion. Conversely, that the eccentric orbits have an inverse ratio to the distances from the sun. Kepler even showed that the orbits are elliptical, so when the planet has completed one-fourth of his whole circuit from the aphelion, it is then found on the half of the distance from the sun, halfway the road between the greatest distance at aphelion and the minimum at perihelion [*Harmonies* 5, 3, 277-278]. The relationships, shown in the *Commentaries on Mars* and in the Book V of the *Epitome of Copernican Astronomy*, namely that the apparent movement is inversely proportional to the squares of the distances from the sun, are not applicable to geocentricism. This is the triumph of the verification of heliocentric theory according to Kepler. To summarize, we observe that the mathematical theory of Kepler calculates the following discrete objects of measurement [Harmonies, 5, 4]: a) distances from the sun, b) periodic times, c) diurnal eccentric arcs, d) daily delays of these arcs, and e) angles at the sun and the diurnal area as it appears to these angles from the sun. All these elements are variable, with reference to the coordinates. As pointed out by Descartes, the variety of properties of sensible bodies are, in the era of scientific revolution, nothing more than specific properties of *size, shape* and *movement*, and indeed, the same is true at all levels, regardless of the level of observation. The same aspirations to widen and generalize the field of mathematical applications we encounter also in the rhetoric of Newton. According to Newton, the stages of young science are three: a) mathematical research of the quantities of forces, b) comparison of these proportions with the phenomena of nature, and c) discussion upon the natural types, the causes and the ratios of forces. It is worth observing that just at this third stage, Newton corrected Kepler. The latter, absorbed in observations, preoccupied in mathematics and learned to favour deductive methods, failed to adequately analyse the physical concepts of *force, attraction, gravity* and did not capture the proper relationship between the forces exerted on celestial bodies and their distance from the sun. This example is characteristic of the complexity of the problems of Mathematization and justifies the great appreciation that Newton held for the inductive methods. It also shows the extent to which astronomical science is bound to deductive schemes, given the distance from the field of observation.

7.1. Space and time as relations

The epistemological rupture of Copernicus, the laws of planetary motions of Kepler, the physical observations of Galileo and Huygens, the early conceptions of relativity, the physical theory of Newton, were components of the extremely fertile and rich in influences cognitive environment, which prompted the restless Leibniz to shape a peculiar theory of space and time, which expressed some of the concerns and intuitions of the scientific community of seventeenth century and in particular the scientific group of the Academy of Sciences of Paris, but stayed until the twentieth, unrealistic.

The problem of the relationship of the mind with space, the study of the nature of the continuous in all mathematical fields, and the observation of the linear and the centrifugal movement, along with the confrontation to atomism, based on the assumption for the absence of vacuum in nature, led Leibniz to the belief that space and time are not real entities but virtual relationships and mathematical concepts. Space is only the order of coexisting phenomena, as time is only the order of successive phenomena. Today we call this view relationalism: spatial and temporal relationships between objects and events are immediate and not reducible to relations of space-time points and all kinds of motion are relational movements of bodies. The starting point of the controversy about the ontological status of space and time was the distinction between primary and secondary qualities developed by Galileo, Descartes, and Locke.

According to Descartes, the primary properties (e.g., height, width, depth), unlike the secondary properties (e.g., roughness, colour), are perceived with an accuracy that depends on the position of the observer to the physical objects. The primary properties reveal the true nature of the body. Descartes, by stating this discrimination, extracts concepts, which Leibniz also analyses: a) the observability, b) the primary, true nature of the body, the object, c) the spatial dimensions, as physical characteristics of the physical object. From the above premises, Newton concludes that: There is a similar distinction between *sensible* space and time (where the sensible secondary properties appear), and the completely differently theorised, *mathematically* defined space and time (where unseen masses and atoms of physics are being placed). Because of that, Newton introduced the terms "absolute and relative, true and apparent, mathematical and common" space, time, motion (*Principia: The Mathematical Principles of Natural Philosophy*, Definition VIII, Scholium). Newton thought that, since we can imagine extramundane space without bodies, and because extension is independent of the bodies, space is certainly not material, but has its own way of existence, without being neither substance nor random. Absolute and relative motion can be identified by their properties, their causes, and their effects. Absolute acceleration developed during rotation and being observable by the diverting forces from the axis is the high spot of the Newtonian argument.

In summary, Newtonian science was distinguishing absolute from relative space and time: Absolute, genuine and mathematical time, in itself and from his nature, flows regularly without dependence on nothing from the exterior, and his other name is continuity. Relative, apparent and common space is a certain perceptible and exterior, detailed and unstable measure of duration of movement. The absolute space, from his own nature, without dependence on anything from outside, remains always the same and motionless. The relative space is a certain mobile dimension or measure of absolute spaces (Gillispie, 1990).

The famous controversy between Leibniz and Newton was not direct but mediated by Clarke, an English theologian. The field of dispute was meant to be that of observability, measurement, and experiment. The relational theory of Leibniz was articulated in juxtaposition to the theory of Newton. Leibniz believed that space is something completely relative, as time; that is to say, space is the order of coexistence, as the time is the order of sequences. Space implies, with terms of possibility, an order of things that exists in the same time, considered as if they exist together, without examining their way of existence. The frame of reference of space and time are the natural objects and their relations, the incidents, the processes. Space is nothing other but the order of existence of things which are observed as they exist together. For the relational theory, the possibility of a material universe as a whole that is moved either in space or in time is strictly without meaning, as long as space and time do not exist independent from the order of bodies and incidents in the universe.

Relationalism and relativity

Leibnizian space was prior to spatial relations, but it was not an absolute framework of measurement. A position in space is determined only in relation to another position, as far as the latter can be regarded as fixed. That is exactly what makes Anapolitanos (1999) to evaluate the theory of Leibniz not only as relationalist but also as relativistic theory since one of the basic postulates of the special theory of relativity is that there is no preferential spatial system of reference. Leibniz, however, remained attached to metaphysical notions such as that of the mind, as the cause of motion. Furthermore, the dominant scientific theory, until the appearance of the theory of relativity, was that space and time were absolute reference systems of things, objects, and events.

The argumentation of Leibniz questioned the verifiability of absolute space because in any systems of coordinates with relatively uniform motion we have not a means to distinguish the absolute uniform motion. "How would the world be, if there was a reportable unobservable change?" "The same," would Leibniz reply. The Newtonian, however, may support the following: it is not true that in any possible world, observability, i.e. observational conditions, disprove the existence of realistic space. Nevertheless, in the special theory of relativity, we still have an absolute motion, in terms of a class of highly abstract and unobservable entities. The speed of light is always the same in all systems of coordinates, whether or not the light source is moving, and with any way it may move (Friedman, 1983; Earman, 1989). How does the speed of light remains constant in two coordinate systems that are in relative uniform motion? This has to do with the relativity of time introduced by the Lorentz transformations.

The contribution of Leibniz in theoretical physics of space and time is operative, though not prescriptive. The emphasis on the concept of relation has contributed indirectly to the discovery of the main results of the theory of relativity, according to the Lorentz transformations: the relativity of simultaneity, the time dilation and the length contraction of the bodies. The special theory of relativity accepts the fixedness of relations only for systems of coordinates that move with relative uniform motion: the time is determined by clocks, the spatial coordinates by sets of rods, the movement may affect clocks and rods, as shown by the effects of the electromagnetic field. Events that in a reference system coincide or occur at the same point in space, in other inertial systems can occur at different times or in different places in space, while the deletion of points of the space-time manifold is used by relativists to construct cosmological models. The motivation of Einstein, when he formulated the general theory of relativity, was, besides the problem of gravity, a thought experiment to eliminate the absolute motion. However, he only revised the concept of absolute space, which was replaced by the concept of the displacement field that is a component of the total field. There is no space without field, space is now a quality of the field. The structures yet are endogenous, the metric of space is a function of the distribution of matter and energy, and the laws of physics are accepted in every system of reference (Jammer, 1993). The relativity of motion presupposes variable structures, frames of reference of the Riemannian geometry. The acceleration here is a result of the curvature of space-time, which produces dramatic changes in the observable gravitational effects. The equivalence of all spatial points, i.e. the homogeneity of space, the isotropy, i.e. the equivalence of all space directions, and the homogeneity of the time, in the light of Leibniz's saying that *a point in space does not differ in any respect whatsoever than any other*, seem to conform to the relationalism, but in fact, the principles of the general theory simply do not apply a preferred status to any class of coordinate systems, any reliance on the principle of inertia and its systems of coordinates.

Questions about the Universe

"The universe in one go", may stand as a current expression for the unified theory. Nevertheless, the unification of the physical theories was implausible before 1865, when James Clerk Maxwell proposed the existence of wavelike disturbances in the combined electromagnetic field; he supported that the waves maintain fixed speeds and that the light is an electromagnetic wave; he also calculated that the radio wavelength is one meter or more, while it is a few centimeters for microwaves, a ten thousand of a centimeter for infrared waves, etc. However, around 1900, the electron theory of matter was proposed, which was incommensurable with the Newtonian mechanical laws. Quantum mechanics had to be developed to resolve the inconsistency (Feynman, 1988). For a very long time, the traditional preference for local action, the ancient belief that nature abhors vacuum, and the related rejection of action at a distance necessitated the supposition of a tangible ether filling all space. Newton's conception of the ether "was a passive background, more like the canvas of a painting hung in a museum than the canvas of a circus tent rippling in the wind" (Wheeler, 1998: p. 326). When Michelson and Morley observed that there was no difference between the speeds of light in the direction of earth's motion and at right angles to the earth's motion, the presupposition of an ether started being undermined. Albeit Lorentz's attempt to support that objects contract and clocks slow down when they move through the ether, Einstein and Poincare showed that the conception of the ether is unnecessary, if one abandons the idea of absolute time. According to the theory of relativity, all observers should measure the same speed of light, regardless how fast they are moving. A remarkable consequence of this idea is the equivalence of mass and energy. Furthermore, there is the law that nothing may travel faster than the speed of light.

Nevertheless, the most significant modern idea in the relativistic theory was the suggestion that gravity is not a force, but a consequence of the fact that space-time is not flat; it is curved, or warped, on account of the distribution of mass and energy in it. After that discovery, generalizations such as the geodesics became essential for the description of the celestial orbits and light rays' paths. Thus, the influence of the mass of the sun to the light rays from a distant star would cause the deflection of the light through a small angle. Furthermore, general relativity predicts that time appears to run slower near a massive body like the earth. These predictions were observationally confirmed in 1919 and 1962, respectively (Kennefick, 2009; Hawking, 1988).

With general relativity, space and time became dynamic quantities; their curvature is being affected by everything that happens in the universe. Space and time are meaningful only inside the limits of the universe. Hence, Einstein's general theory of relativity implies, as Hawking (1988) supported, that the universe may have a beginning, and, probably, an end. Another question refers to the relative positions and the distances to the distant, remote and 'fixed' stars. The nearest star is Proxima Centauri, about four light years away, whereas many other visible stars lie within a few hundred light-years of earth. Remarkable was Edwin Hubble's demonstration that ours is not the only galaxy; and Edwin Hubble's and Georges Lemaître's observations that everywhere in the visible universe, distant galaxies are moving rapidly away from us (Reich, 2011). Therefore, the universe is expanding, as Alexander Friedmann insisted (Tropp et al. 1993), probably beginning with a Bing-

Bang. Currently, we know the existence of some hundred thousand
million galaxies, each one containing some hundred thousand
million stars. Our galaxy "is about one hundred thousand light-
years across and is slowly rotating" (Hawking, 1988: p. 39). The
color of the light, the different spectrum, is the sole characteristic
feature for the vast majority of the stars. Temperature and thermal
spectrum are crucial for the light emitted by any opaque object
glowing red hot. Furthermore, the Doppler Effect is a realistic confi-
guration of the aforementioned differences in the thermal spectrum:
Stars and quasars moving away from us have their spectra redshifted,
while stars moving toward us have their spectra blue shifted.

> When the source is approaching the observer, the wave
> crests 'bunch up,' resulting in a shorter observed wave-
> length; this is a blueshift. When the source is receding
> from the observer, successive wave crests are 'stretched,'
> giving a longer observed wavelength; this is a redshift.
> When viewed from the transverse direction (i.e., perpendi-
> cular to the direction of motion), the wavelength is un-
> changed...
>
> The classical Doppler Effect occurs with relative mo-
> tion because the crests of the light waves bunch to-
> gether (for relative approach) or stretch out (for rela-
> tive recession). Relativity adds to this effect a correc-
> tion: the frequency of the light, which is an inverse
> time interval, is less at the source than at the receiver,
> due to time dilation. (Hawley and Holcomb, 1998: p.
> 96, 183).

Hubble's discovery was a result of the observation that nearly all
galaxies appeared red-shifted, while the galaxy's redshift is directly
proportional to the galaxy's distance from us. Even Einstein, however, when he introduced the general theory in 1915, contended
that the universe is static. This is why, he needed the so-called cosmological constant, which meant to be an 'antigravity' force, an
inbuilt tendency of the space-time to expand, balancing, thus, the
attraction of the matter (Topper, 2015). A solid empirical corroboration of the general theory may be provided by observational facts,
such as the anomalous precession of the perihelion of the planet
Mercury, the bending of light in gravitational fields, the deflection

of light by the sun, and the precision tests that started after 1959 (Einstein, 1916). Another area of questions is related to recent discoveries, such as the cosmic microwave background radiation; resulting from computations of wave frequencies in the case of light, which range from four to seven hundred million waves per second, while in microwaves ranging only ten thousand million waves per second (Hawking, 1988).

Moreover, the existence of the vacuum,[12] which was approached experimentally after the invention of the air pump, relates to a conundrum of modern Cosmology, created by the adoption of the idea of a Black Hole, as the center of the attraction of a galaxy. The Black Holes theories are usually consisted of a Bing-Bang singularity core, frequently questioned by the scientific determinists. Further discussions are caused by the efforts for a geometrization (of magnetic and gravitational fields) and a unification of the relevant theories, e.g. gauge theories (Goenner, 2004). Therefore, with the addition of the Uncertainty Principle and Quantum Mechanics, the finiteness of the Universe and the boundaries of it, have to be reconsidered, in their connection with the relativistic theoretical conceptions, such as the strong and weak equivalence principle (Rabinowitz, 2007). On the opposite direction, scientists as Thomas Gold rejected the Bing-Bang theory, supporting an infinite conception of time. The fruitful theory production notwithstanding, success was limited, because of the lack of a unified theory, in other words, the need to apply a variety of theories at different phenomena of the physical world. The quantum electrodynamics (QED), for instance, the theory that describes how light and matter interact, has no application to gravitational and radioactivity phenomena. QED explains mainly phenomena behind chemistry: "Gasoline burning in automobiles, foam and bubbles, the hardness of salt or copper, the stiffness of steel" (Feynman, 1988: p. 8). For the theoretical unification of all the explanations of the physical phenomena one must call upon the intervention of the philosophy of science.

---

[12] The idea of the vacuum was rejected by the mathematician Leibniz (1969; *G. W. Leibniz and Samuel Clarke: Correspondence*, 2000: V, 34).

Chapter 8

Philosophy and History of Science

From Galileo Galilei to Edwin Hubble, natural science dared a breathtaking ascent, based on Mathematization, on the practical turn of the sciences, on technical inventions, scientific instruments, experiment, and specialization. The role of history and philosophy of science was outstandingly critical in these developments. Philosophy of Science aims to conceive and analyze the processes of scientific conceptualization, justification, research, ethics and discovery. History of Science studies the masterpieces of scientific advancement and explores the continuities and the ruptures, outlining the characteristics of the historical periods, the worldwide dissemination of science and the scientific revolution.

The meaning of facts and theories and the relationship between facts and theories has been proposed as a definition of the philosophy of science. One of the best examples of this kind is 'The Genesis and Development of a Scientific Fact,' the most famous book of the immunologist Ludwig Fleck. The scientific facts, according to Fleck (1979), are produced by styles of thought and thought collectives, which are based on directed perception and vision.

An elemental task of the philosophy of science is the definition of certain scientific disciplines, such as the 'high energy physics,' which is an in-depth scientific research of the matter in extremely small distances, in contradiction with the in-breadth inquiry of all other physical disciplines. Furthermore, the definition of the notions "cause" and "effect" is one of the well-known instances of significant epistemological debates, while in modern times determinism is threatened by quantum mechanics. On the whole, the mission of philosophy, regarding the approach to the history of science, aims at the clarification of the basic scientific conceptual systems, laws, explanations, and theories; for example, the need to explain the theory of quantum chromodynamics, namely the theory of strong nuclear force. The inherent advancement of particle physics would remain totally obscure without a rational analysis of the genuine contributions of Max Planck, Albert Einstein, Ernest Rutherford, Niels Bohr, Ernest Schrödinger, Werner Heisenberg, Wolfgang Pauli, Louis de Broglie, Paul Dirac, Hideki Yukawa, Enrico Fermi, C.N. Yang, Robert Mills, Gerard 't Hooft, Martinus Veltman, to the elaboration of the initial theory (Maerivoet, 2001).

According to sceptics, nominalists, and conventionalists, such as Ernst Mach (1883), scientific laws are equivalent to artificial creations. Henri Poincaré insisted, however, that industrial applications and technological systems, reveal a realism of "relations", indications of connections that are grasped in approximate laws.

From another viewpoint, there is an epistemological rupture between everyday life and scientific theory, as Bachelard supported. The technological applications formulate new realities, which could be approached with artificial conceptions, such as the Phenomeno-technique, namely the essential contribution of technology to the contemporary scientific modus operandi. As modern science invents new phenomena, nothing is given, everything is constructed (Rheinberger, 2005).

> *This is exactly what happened in the Weinberg-Salam model. In that theory, the mass of the electron is not intrinsic; it comes instead from its interaction with certain other particles, which are called Higgs particles. If there were no Higgs particles, the electron would have no mass. It would move at the speed of light, like a photon. But if it finds itself surrounded by a gas of Higgs particles, an electron is not able to move so quickly* (Smolin, 1997: p. 54)

The epistemological tasks of the philosophy are the study of knowledge and justification, the defining components of knowledge, the substantive conditions and resources of knowledge, the limits of knowledge (Moser, 2002). Philosophy of science, therefore, may generalize, interrogate and refine the scientific findings, on account of their credibility, interconnection, applicability, scope, interpretation, valuation etc. The inquiry on the prerequisites for the confirmation of a theory has found many attractive fields of implementation, such as in the use of fossil records, homologies and isomorphisms between organisms, climatic, geographical and environmental factors in the confirmation of the Darwinian theory of natural selection. Justification, truth and belief are the individually necessary and jointly sufficient components of propositional knowledge, in order to obtain adequately indicated knowledge, evidence for a "justified true belief". Regarding truth, different accounts of verification, corroboration, deduction, induction, and falsification have significant implications for scientific practice (Lakatos, 1968; Hempel, 2001; Popper, 2010). However, justification is not the one and only source of truth. Discovery, the reality of the laboratory, as "the other side of the moon", is an equally significant scientific process. In other words, history of science has a didactic role, namely the use of the historical method for the instruction of science lessons, as Pierre Duhem proposed. History of science may research and exhibit the styles of thought and the logical structures of scientific theories. For this reason, a main problem in the philosophy of science is the generation of general scientific principles out of the critical examination of common-sense statements, an interpretation that leads to scientific propositions, for instance from the common-sense of 'sluggishness' to the scientific concept of 'inertia'. Another task of the philosophy of science is to find a theory that carries out satisfactorily sensible observations.

Even the most hard-boiled engineers must recognize that there are two types of statements: on the one hand, statements regarding direct observations and crudely empirical rules which the engineer calls "rules of thumb"; on the other hand, intelligible principles like the law of inertia. No one can deny that these two levels exist. One of the most obvious differences between these two levels is this: The engineer will readily change his "rules of thumb" under the impact of new observations, but he will not easily admit that such a general principle as the law of inertia is wrong. If it comes to a choice, he will usually assume that his observations were wrong and not the law of inertia (Frank, 1957: p. 11).

Nevertheless, the philosopher Thomas Kuhn (1977; 1981; 1996) has shown how science through paradigm shifts and epistemic experiences of incommensurability, undergoes a discontinuous, non-rational development. Simultaneously, the philosophy is excessively important for our ethical, moral approach to science, as it was proved by the cases of the philosopher Immanuel Kant and the works of the physicists Enrico Fermi, Lise Meitner, Otto Hahn, Fritz Strassman, Leo Szilard and Eugene Wigner. Ethical elements may

be also uncovered in the worldview of the philosopher Paul Feyerabend (1993), who criticised rationalism, ideology, and government
in the name of freedom, introducing a self-standing existence of
science, independent of systems of dogmatic statements.

8.1. Continuities, ruptures, and transitions

Truth opens slowly, according to Newton, like the first dawning, by
little and little into *full and clear light.* By the emergence of early
modern science in the Renaissance, the divorce from philosophy
had not yet been completed in the circles of scientific communities.
In fact, philosophy and science had to define themselves and their
relationship under numerous, successive breakings, from the birth
of geometry, by the third century BC, to the emergence of biology
and psychology in the nineteenth century.

The birth of the new sciences, for instance life sciences, yields
leading-edge tasks for the philosophical inquiry, as it was shown by
the failure of atomism and mechanics to explain the phenomena of
the life sciences. Philosophical research had to stand up against
common-sense in its acquaintance with the Newtonian concepts of
mass and force, with the relativistic notions of spatial and temporal
realities, with the revision of common conceptions of rest and motion, velocity and position, cause and effect, freedom and determinism by the Quantum Theory. The non-Euclidean geometries, which
contributed to the emergence of the theories of relativity and quantum physics, constitute an important philosophical theme in the
history (Rosenfeld, 1988) and the philosophy of science. Moreover,
an epistemological shock in the Quantum Mechanics related to the
ambivalence at the interaction between observers and observed.
Therefore, history of science includes breakdowns that undermine
the systematic objectives of science and philosophy.

The allegedly orderly structured domains of science and
philosophy, presupposed simultaneously a variety of disrupting
discontinuities, transitions, rejections, contradictions and transformations during their historical development. From the beginning of modern times, the *Accademia del Cimento at Florence*
(founded in 1657), the *Royal Society of London* (founded in 1662),
and the *Académie des Sciences* in Paris (founded in 1666) played a
vital role for the dissemination and the secularization of science. In
the same time, preconditions of the scientific revolution were the
advancements in analysis and algebra, the turn to the natural and
applied sciences, and the wider use of scientific instruments and

experiments, while simplicity was emphasised as an additional requirement in scientific research.

The dissemination of the use of coins, the invention of letters of credit, bills of exchange, accounting and bookkeeping gave rise to financial and banking enterprises. The generation of the numeric system and its adaptation to the commercial practices emanates from the works of the abbacists, according to van der Waerden (1985), and reflects the prominence of the integration of mathematical practice to the *scheduled* development of the community. This demand was also reflected in the works of François Viète and Dmitri I. Mendeleyev, namely in their provision for void places in the construction of the symbolic patterns of their respective sciences.

Meanwhile, J.B. Richter (1762-1807) had tried to determine "the weights of various alkaline substances (potash, soda, magnesia, lime) which were neutralised by a given quantity of acid (hydrochloric acid, sulphuric acid)" (Hooykaas, 1999: p. 53), whereas the homologous series of carbon compounds, such as methyl, aethyl, glyceryl, amyl, offered a more fruitful instance of Mathematisation. The possibility to synthesize the missing organic elements of the proposed series, created another challenging theme of philosophical research, regarding the differences between organic and inorganic chemical science.

The quest for the *lapis philosophorum*, for refining agents, seeds and ferments presupposed a productive process: The old story of the sons who, on the basis of a saying of their dying father, dug up a vineyard for a hidden treasure, and they did not find what they were seeking, nevertheless they were indirectly enriched by the fact that their work had raised the fertility of the soil, finds a complete parallel in the history of the development of chemistry. The illusion of the alchemists to make gold has prompted the researchers for centuries to subject the substances which they encountered in nature to various physical and chemical treatments; and from the experience so gained something has finally emerged that is more valuable than gold: modern chemistry (Dijksterhuis, 1983: p. 92).

Mathematised scientific practice was extremely significant not only in terms of measurement and construction, such as with chemistry, astronomy, and optics, but also with respect to the verification and falsification processes. More precisely, the falsification of the Aristotelian doctrines, such as that the velocity of falling bodies varies with their weight, indicated that Galileo was serving nothing less than a new science.

The overcoming of Aristotelianism occupied not only the modern science of mechanics but also astronomers (Copernicus) and chemists (Jungius, Boyle). Moreover, during the seventeenth century, the microscope, the telescope, the thermometer, the barometer, the air-pump, and the pendulum clock were invented. According to Heilbron (1990), the scientific instruments may be classified to measurers (such as the barometer, the thermometer, the calorimeter, the electrometer and the chemical balance), explorers (such as the air pump and the electrical machine) and finders (telescopes, chronometers, theodolites). Simultaneously, new theoretical paradigms were emerging on account of the evidence collected by the scientific instruments. For instance, heliocentrism was corroborated by observations made by Galileo with the telescope (Sunspots, Jupiter's satellites, Venus' phases, the nature of the Milky Way, Saturn's Rings etc.).

Indirect evidence against Aristotelian views was firstly attained in 1604, by the observation of a new star, which according to the Aristotelians, should be located within the lunar sphere, because in the outer sphere no change could appear. In 1609, the dauntless Galileo constructed a telescope on the Dutch pattern and used it as a scientific instrument, to discover the "Medicean stars", the four satellites of Jupiter, mountains and valleys on the Moon, forty fixed stars in the constellation Pleiades, nebulae etc.

Mathematization was extremely significant not only in terms of measurement and construction, such as with astronomy and optics but also with respect to the *verification* and *falsification* processes. A point that has to be underlined is the pertinence of the scientific cooperation, which became manifest by the fact that not only Galileo, but also Kepler with the naked eye, and Fabricius and Scheiner through their telescopes, had observed the Sun-spots. Moreover, Simon Marius in 1612 may have preceded Galileo in the observation of a nebula in the constellation Andromeda.

Concerning the philosophy of science, two significant groundbreaking issues in the Galilean turn were: a) the abandonment of the Aristotelian belief in the indestructibleness of the heavenly bodies, b) the reversal of the Aristotelian opinion for the physical position of a terrestrial body. Whereas Aristotle explained gravity with the tendency of a terrestrial body to remain in its physical position, namely the state of rest on the earth, Galileo insisted that the physical stance of a body was motion.

Inertial frames became important after the Copernican turn, simply because the earth moves along with every terrestrial body. The *Dialogue* and other pro-Copernican works, however, remained on the Index until 1822, whereas only in 1838 astronomical instruments and methods could sufficiently measure the stellar parallax (F. W. Bessel).

Conclusions

The advancements of technology, the corresponding progress in research methodology, and their impact in scientific discoveries is one of the more interesting themes in philosophy of science. In the midst of the seventeenth century, the scientists Leeuwenhoek and Huygens were investigating lens making and Robert Hooke was writing his *Micrographia* and *An Attempt to prove the Motion of the Earth from Observations*. In the eighteenth century, the quantifying spirit was widely disseminating and becoming dominant in science, altogether with a dramatic increase in the precision of the instruments of physical science (Frängsmyr et al. 1990). Robert Hooke's works are representative instances of visual culture in science. Hooke, as a constructor of Gregorian telescopes and microscopes, begins his *Micrographia* with an impressive graphical representation of his scientific instruments. The *Attempt* begins with the proposal to use an *experimentum crucis*, based on the problem of stellar parallax, in order to decide between the Tychonian and the Copernican hypotheses.

Apart from the well-studied paradigms of physics, chemistry and computing, biology and geology may offer a fertile ground for prospective philosophical researches in the history of science. Sixty million years ago, the so-called *Cambrian explosion* was characterised by the emergence of sophisticated forms of life. The philosophical analysis of the notion of life, the related mutability of species and the one-tree-of-life hypothesis required yet an appropriate methodology. The emergence of a visual language, with the use of maps, shaped the young science of geology in the nineteenth century. For such reasons, Hans Reichenbach in his book *The Rise of Scientific Philosophy* contended that there is a mathematical vision and an eye of the mind, which studies objects similar to the platonic Idea. The image of a vortex, for example, is indwelling during the long seventeenth century not only in the experiments of Newton and Huygens but also in the physical worldview of Descartes.

The most astounding implication created by this visual way of theorising was its alleged confirmation through modern astrophotography. Semantics, however, may not have always visual form. In the search for generalisation, Charles Sanders Peirce suggested a triadic division of the scientific object to: a) semantical, the relations between physical object and signs, b) logical, the relations between symbols, and c) pragmatic component of science, between the scientist and his signs, the social and psychological circumstances. Philosophy of science may also delineate the presuppositions of the communication between common sense and scientific knowledge.

From this perspective, science, technology, and philosophy interrelate to each other in many aspects, as in the domains of sociobiology, biotechnology, geo- and bioinformatics, visual humanities etc. In parallel, the role of invention and innovation is increasingly powerful in modern societies. George Eastman, the inventor of the camera, as he was the sole user of it in the beginning, he had also to define the target group of his invention: Who might be the users of the camera? The users of technology may be agents of technological change and active participants in the social construction of technology. Other influential modern approaches may focus on the system builders, who construct the large technological systems or emphasize the role of actants (humans, machines and natural forces), in the frame of the actor-network theory. The sociology of scientific knowledge, altogether with the disciplines of science policy and science communication, created growing interest for the actual practices of the scientists and the scientific instruments.

Continuities, ruptures, and transitions in the history of science and technology are realized through a transition from the culture of science to a culture of research, according to Latour (1998). Science is certainty; Research is uncertainty. Science is cold, straightforward and detached; Research is warm, provoking and risky. The traditional science was supposedly at the core of the society. Suddenly, the anthropocentric, socio-centric, logocentric philosophy of science was hopelessly bogged down, because its Kantian standards were moulding in a one-dimensional tug of war between nature and society. Undoubtedly yet, in modern industrial times, the scientific and technological advances were cooperatively organised through systematic main processes: classification, design, measurement, experiment, and construction. The theoretical source of innovation remained increasingly critical in these transformations, such as between the rival mathematical and technical versions of aviation engineering (Bloor, 2011). Although the development was gradual from the ancient to classic and modern culture, we can nevertheless point out essential breakthroughs, transitions and ruptures, for instance the Pythagorean view that things are numbers, the atomist conception of a corpuscular nature, the invention of the concept of energy, Leonardo's mechanics, Boyle's atomism, mechanical philosophy and natural theology, Galileo's kinetics, shipbuilding, industrialization, railways, electrification, high-temperature superconductors and beyond.

During this long historical process, philosophy of science inspired ingenious attempts to overcome established clichés, such as with the introduction of the discontinuity of energy with the notion of quanta by Max Planck in 1900; the attempt to transcend the incongruence between classical electrodynamics and Rutherford's atomic theory, which triggered the modern approach of Bohr's quantum mechanics (namely the postulation, against Maxwell's theory, that light would be emitted only when electrons jumped from a higher to a lower energy level), and, furthermore, the emergence of Schrödinger's wave mechanics, Heisenberg's matrix mechanics, and the so-called *Copenhagen Interpretation* (Murdoch, 2001). In conclusion, the future philosophers of science may expect long debates on recurring philosophical problems, such as determinism, although it was rejected by the *energy-time* principle of uncertainty in quantum mechanics. The so-called *Einstein-Bohr Debate*, connected with the questioning of the possibility of dice-playing in nature (Bokulich, 2008), has been a fruitful challenge for the creation of new research projects and theoretical attempts. It became evident that the reveal of modern theoretical notions such that of the quantum nonlocality, along with the recognition that certain physical theories are radically incomplete, offer fertile ground for the philosophy of science. This is why, the communication and cooperation between diverse sciences, such as quantum mechanics, physics and chemistry, is an indispensable responsibility of the philosophy of science and probably its most significant task.

Bibliography

Abdalla, M. (2007). Ibn Khaldun on the Fate of Islamic Science after the 11th Century. *Islam and Science,* 5(1): pp. 61-70.

Aczel, Amir D. (1999). *God's Equation: Einstein, Relativity, and the Expanding Universe.* New York: Four Walls Eight Windows.

Adams, J. (1996). Principals and Agents, Colonialists and Company Men: The Decay of Colonial Control in the Dutch East Indies. *American Sociological Review,* 61(1): pp. 12-28.

Agricola, Georgius (1950). *De Re Metallica.* New York: Dover.

Alexander, H. G. (Ed.), (1956). *The Leibniz-Clarke Correspondence.* Manchester University Press.

Al-Hassan, Ahmad Y. and Donald R. Hill (1986). *Islamic Technology: An Illustrated History.* Cambridge University Press.

Allen, R.E. (1959). Anamnesis in Plato's 'Meno' and 'Phaedo'. *The review of metaphysics,* 13 (1): pp. 165-174.

Allen, Richard B. (2010). Satisfying the 'Want for Labouring People': European Slave Trading in the Indian Ocean, 1500–1850. *Journal of World History* 21(1): pp. 45-73.

Anapolitanos, D. (1999). *Leibniz: Representation, Continuity and the Spatiotemporality.* Dordrecht: Kluwer Academic Publishers.

Anderson, Benedict (2006). *Imagined Communities: Reflections on the Origin and Spread of Nationalism* (3rd Ed.). London: Verso.

André-Julien Fabre (2008). Pneumatic Machines in Antiquity (Air as Source of Energy in the *Treatise on Pneumatics* of Heron of Alexandria). *Analecta Historico Medica,* VI (1): pp. 67-70.

Anyanwu, Chika J. (1998). Virtual World and Virtual Reality. *Journal of Australian Studies* 154.

Appadurai, Arjun (2005). *Modernity at Large: Cultural Dimensions of Globalization* (7th Ed.). Minneapolis, London: University of Minnesota Press.

ARAMCO withdrawn from talks to buy the country's state-owned refinery shares (1997). *Oil,* 130: p. 3.

Ariew, R. (Edited, with Introduction). (2000). *G. W. Leibniz and Samuel Clarke: Correspondence.* Indianapolis/Cambridge: Hackett.

Aristotle (1938). *The Constitution of Athens; Poetics.* Athens: Papyrus.

———— (1939). *On the Heavens.* Loeb Classical Library. Harvard University Press.

———— (1960). *On Interpretation.* Opera. Berolini apud W. De Gruyter et socios: 16-24.

———— (1960). *On Sophistic Refutations.* Opera. Berolini apud W. De Gruyter et socios: pp.164-184.

———— (1960). *On the Soul.* Athens: Papyrus.

———— (1960). *Posterior Analytics.* Opera. Berolini apud W. De Gruyter et socios: pp. 71-100.

———— (1960). *Topica.* Opera. Berolini apud W. De Gruyter et socios: pp. 100-164.

———— (1980). *Poetics*; 'Longinus,' *On the Sublime*; Demetrius, *On Style.* London: The Loeb Classical Library.

———— (1993). *Politics.* Athens: Cactus.

———— (1997). *Physics.* Athens: Cactus.

Ashtekar, A. (2006). Space and Time: From Antiquity to Einstein and Beyond. *Resonance* 11 (9): pp. 4-19.

Audi, R. (2003). *Epistemology: A Contemporary Introduction to the Theory of Knowledge* (2nd Ed.). New York: Routledge.

Ayres, Robert U. (1990). Technological Transformations and Long Waves. Part I. *Technological Forecasting and Social Change* 37: pp. 1-37.

———— (1997). *Industrial Metabolism: Work in Progress.* Working Paper 97/09/EPS, INSEAD, Fontainebleau, France.

Babbage, Charles (1864). Of the Analytical Engine. In: *Passages from the Life of a Philosopher*, Ch. VIII. London: Longman, Green, Longman, Roberts, & Green.

———— (1889). *Babbage's Calculating Engines Being a Collection of Papers Relating to Them; Their History and Construction.* Edited by Henry P. Babbage. London.

———— (2009; first published 1832). *On the Economy of Machinery and Manufactures.* Cambridge: University Press.

Bacon, Francis (1989). *The Works of Francis Bacon* (ed. James Spedding, Robert Leslie Ellis, and Douglas Denon Heath) 14 vols. London, 1857–74; facsimile reprint, Stuttgart/Bad Cannstatt.

Baird, D. (2004). *Thing Knowledge: A Philosophy of Scientific Instruments.* Berkeley, CA: University of California Press.

Ballard, K. E. (1960). Leibniz's theory of Space and Time. *Journal of the History of Ideas* 21: pp. 49-64.

Barker, Peter (2001). Kuhn, Incommensurability, and Cognitive Science. *Perspectives on Science* 9(4): pp. 433-462.

Barrera-Osorio, Antonio (2006). *Experiencing Nature: The Spanish American Empire and the Early Scientific Revolution.* Austin: University of Texas Press.

Beamish, Anne (2008). *Learning from Work: Designing Organizations for Learning and Communication.* Stanford, CA: Stanford Business Books.

Beazley, C.R. (1895). *Prince Henry the Navigator, the Hero of Portugal and of Modern Discovery, 1394-1460 A.D. With an Account of Geographical Progress throughout the Middle Ages as the Preparation for His Work.* New York: G. P. Putnam's Sons.

Bell, David (2001). *An Introduction to Cybercultures.* London: Routledge.

————— (2006). *Science, Technology and Culture.* Berkshire: Open University Press.

Ben-David, Joseph (1971). *The Scientist's Role in Society: A Comparative Study.* Prentice Hall.

Bengtson, Herman (1991). *History of Ancient Greece* (A. Gavrelis, Trans.). Athens: Melissa.

Benson, H.H. (2003). The Method of Hypothesis in the 'Meno'. *Proceedings of BACAP* 18, 2003, pp. 95-126.

Berdayes, Vicente, and John W. Murphy (Eds.). (2000). *Computers, Human Interaction, and Organizations: Critical Issues.* Westport, CT: Praeger.

Berlin, Leslie (2005). *The Man behind the Microchip: Robert Noyce and the Invention of Silicon Valley.* New York: Oxford University Press.

Bernard, Alain (2003). Ancient Rhetoric and Greek Mathematics: A Response to a Modern Historiographical Dilemma. *Science in Context* 16 (5): pp. 391-412.

————— (2003). Comment définir la nature des textes mathématiques de l'antiquité grecque tardive? Proposition de réforme de la notion de «textes deutéronomiques». *Revue d'histoire des mathématiques* 9: pp. 131-173.

Bernard, Claude (1865; 1957). *An Introduction to the Study of Experimental Medicine* (H.C. Greene, Trans. 1927). New York: Dover.

Berraha, N., J. Bozekb, J.T. Costelloc, et al. (2010). Non-linear processes in the interaction of atoms and molecules with intense EUV and X-ray fields from SASE free electron lasers (FELs). *Journal of Modern Optics* 57(12): pp. 1015-1040.

Berthold, A. (2011). Die Darstellung von Raum auf griechischen Münzen. *eTopoi,* 1: pp. 69–98.

Bertoloni Meli, Domenico (1993). *Equivalence and Priority, Newton versus Leibniz, including Leibniz's unpublished manuscripts on the Principia.* Oxford: Clarendon Press

Biagioli, Mario (2003). Stress in the Book of Nature: the Supplemental Logic of Galileo's Realism. *MLN* 118(3): pp. 557-585.

Bikson, Tora K., and Constantijn W. A. Panis (1999). *Citizens, Computers, and Connectivity: A Review of Trends.* Santa Monica, CA: Rand.

Bintliff, J. (2012). 'The Immense Respiration of a Social Structure:' An Integrated Approach to the Landscape Archaeology of the Mediterranean Lands. *eTopoi,* Special Volume 3: pp. 1–9.

Bird, Richard J. (2003). *Chaos and Life: Complexity and Order in Evolution and Thought.* New York: Columbia University Press.

Biringuccio, Vanoccio (1943). *Pirotechnia.* New York: The American Institute of Mining and Metallurgical Engineers.

Birkhoff, Garrett and John Von Neumann (1936). The Logic of Quantum Mechanics. *Annals of Mathematics,* Second Series, 37(4): pp. 823-843.

Blackmore, J. (2002). *Manifest Perdition: Shipwreck Narrative and the Disruption of Empire.* Minneapolis: University of Minnesota Press.

Bloor, David (1991). Knowledge and Social Imagery (2nd Ed.). Chicago: University of Chicago Press.

————— (2011). *The Enigma of the Aerofoil. Rival Theories in Aerodynamics, 1909–1930.* Chicago and London: The University of the Chicago Press.

Bokulich, Alisa (2008). Paul Dirac and the Einstein-Bohr Debate. *Perspectives on Science* 16(1): pp. 103-113.

Bolter, Jay David (2000). *Writing Space: Computers, Hypertext, and the Remediation of Print,* 2nd ed. Mahwah, NJ: Lawrence Erlbaum Associates.

Bose, Sugata (2006). *A Hundred Horizons: The Indian Ocean in the Age of Global Empire.* Cambridge, Mass.: Harvard University Press.

Bowker, Geoffrey C., Susan Leigh Star, William Turner, and Les Gasser (Eds.). (1997). *Social Science, Technical Systems, and Cooperative Work: Beyond the Great Divide.* Mahwah, NJ: Lawrence Erlbaum Associates.

Boxer, C.R. (1959). *The tragic history of the sea, 1589-1622: Narratives of the shipwrecks of the Portuguese East Indiamen Sao Thome (1589), Santo Alberto (1593), Sao Joao Baptista (1622), and the journeys of the survivors in South East Africa.* Cambridge, England: Hakluyt Society.

Brazil: Delaying the sale of a 35% stake of Petrompras by the government. State company the Petrompras, as a result of political conflicts (2000). *Oil,* 140: p. 2.

Briar, Celia (1997). *Working for Women? Gendered Work and Welfare Policies in Twentieth-Century Britain.* London: UCL Press.

Brinley, Thomas (1993). *The Industrial Revolution and the Atlantic Economy: Selected Essays.* New York: Routledge.

Broadbent, Jane, Michael Dietrich, and Jennifer Roberts (Eds.). (1997). *The End of the Professions? The Restructuring of Professional Work.* London: Routledge.

Bromley, Allan G. (1987). The Evolution of Babbage's Calculating Engines. In: *Annals of the History of Computing* 9(2): pp. 113-136.

Brorson, Stig (2006). The Seeds and the Worms. Ludwig Fleck and the Early History of Germ Theories. *Perspectives in the History of Biology and Medicine* 64(1): pp. 64-76.

Brown, Clair, Yoshifumi Nakata, Michael Reich, and Lloyd Ulman (1997). *Work and Pay in the United States and Japan.* New York: Oxford University Press.

Buchanan, Brenda J. (2006). *Gunpowder, explosives and the state: a technological history.* Aldershot: Ashgate.

Büchel, Bettina S. T. (2001). *Using Communication Technology: Creating Knowledge Organizations.* New York: Palgrave.

Bullinger, Hans-Jörg, and Jürgen Ziegler (Eds.). (1999). *Human-Computer Interaction: Communication, Cooperation, and Application Design.* Mahwah, NJ: Lawrence Erlbaum Associates.

Butorac Marc (2001). *From The Other Oil Field: Mendeleev, the West and the Russian Oil Industry.* Doctoral Thesis, McGill University, 2001.

Cain, Peter J., and Antony G. Hopkins (2001). *British Imperialism, 1688-2000* (2nd ed.). Harlow and New York: Macmillan.

Camino, Mercedes M. (2005). *Producing the Pacific: Maps and narratives of Spanish exploration (1567-1606).* Amsterdam: Rodopi.

Campbell, Gwyn (Ed.). (2005). *Abolition and Its Aftermath in Indian Ocean Africa and Asia.* London: Routledge.

Campbell-Kelly, Martin, and William Aspray (2014). *Computer: A History of the Information Machine* (3rd Ed.). Boulder, CO: Westview Press, 2014.

Carter, Matt (2007). *Minds and Computers: An Introduction to the Philosophy of Artificial Intelligence.* Edinburgh: Edinburgh University Press.

Chandler, Alfred D., Jr. (1962). *Strategy and Structure: Chapters in the History of the Industrial Enterprise.* Cambridge, MA: M.I.T. Press.

———— (1999). *The Visible Hand: The Managerial Revolution in American Business* (15th ed.). Cambridge, MA: The Belknap Press of Harvard University Press.

———— (2005). *Inventing the Electronic Century: The Epic Story of the Consumer Electronics and Computer Industries.* Cambridge, MA: Harvard University Press.

Chandler, Alfred D., Jr., and James W. Cortada (Eds.). (2000). *A Nation Transformed by Information: How Information Has Shaped the United States from Colonial Times to the Present.* New York: Oxford University Press.

Chandler, Alfred D., Peter Hagstrom, and Orjan Sölvell (Eds.). (1999). *The Dynamic Firm: The Role of Technology, Strategy, Organization, and Regions.* Oxford, England: Oxford University Press.

Chatalian, G. (1991). *Epistemology and Skepticism: An Enquiry into the Nature of Epistemology.* Carbondale, IL: Southern Illinois University Press.

Christianidis, J. (2007). The way of Diophantus: Some clarifications on Diophantus' method of solution. *Historia Mathematica* 34: pp. 289–305.

Cifoletti, Giovanna C. (1995). La *Question* de l'Algèbre: Mathématiques et Rhétorique des Hommes de Droit dans la France du 16ᵉ siècle. *Annales. Histoire, Sciences Sociales* 50(6): pp. 1385-1416.

Clark, Andy (2003). *Natural-Born Cyborgs: Minds, Technologies, and the Future of Human Intelligence.* New York: Oxford University Press.

Clulow, Adam (2006). Pirating in the Shogun's Waters: The Dutch East India Company and the *San Antonio* Incident. *Bulletin of Portuguese-Japanese Studies* 13: pp. 65-80.

Cognitive Science Society (U.S.). (1996). *Proceedings of the Eighteenth Annual Conference of the Cognitive Science Society,* July 12-15, 1996, University of California, San Diego. Edited by Garrison W. Cottrell. Mahwah, NJ: Lawrence Erlbaum Associates.

Cohen, Floris H. (1994). *The Scientific Revolution: A Historiographical Inquiry.* Chicago: The University of Chicago Press.

Cohen, Paul A. (2003). *China Unbound: Evolving Perspectives on the Chinese Past.* London, New York: Routledge.

Cole S. A. (1996). 'Which Came First, the Fossil or the Fuel?' *Social Studies of Science* 26: pp. 733–66.

Cole, S. A. (1998). 'It's a Gas!' *Lingua Franca,* December 1997/January, 1998: pp. 11–13.

Collier, James Lincoln (2004). *Gunpowder and Weaponry.* New York: Benchmark Books.

Collins Harry and Trevor Pinch (1994). The world according to Gold: disputes about the origins of oil. In: Harry Collins and Trevor Pinch, *The Golem: What Everyone Should Know About Science* (Cambridge: Cambridge University Press): pp. 72-92.

Collis, Betty A., Gerald A. Knezek, Kwok-Wing Lai, Keiko T. Miyashita, Willem J. Pelgrum Tjeerd, and Takashi Sakamoto (1996). *Children and Computers in School.* Mahwah, NJ: Lawrence Erlbaum Associates.

Colombia: Danger to become the country importer of oil from oil-producer that is today (2000). *Oil,* 140: p. 3.

Comstock, Helen (Ed.). (1958). *The Concise Encyclopedia of American Antiques.* New York: Hawthorn Books.

Conee, E., and R. Feldman (2004). *Evidentialism: Essays in Epistemology.* Oxford: Clarendon Press.

Conlin, David L., and Larry E. Murphy (2002). Shipwrecks. In C.E. Orser Jr. (ed.), *Encyclopedia of Historical Archaeology.* London and New York: Routledge: pp. 500-502.

Conway, Flo, and Jim Siegelman (2005). *Dark Hero of the Information Age: In Search of Norbert Wiener the Father of Cybernetics.* New York: Basic Books.

Cook, Daniel J., and Henry J. Rosemont (1981). The Pre-Established Harmony between Leibniz and Chinese Thought. *Journal of the History of Ideas* 42(2): pp. 253-267.

Cook, Weston F. Jr. (1994). The *Hundred Years War for Morocco: Gunpowder and the Military Revolution in the Early Modern Muslim World.* Boulder, CO: Westview Press.

Cooper, Joel, and Kimberlee D. Weaver (2003). *Gender and Computers: Understanding the Digital Divide.* Mahwah, NJ: Lawrence Erlbaum Associates.

Copeland, B. Jack, ed. (2006). *Colossus: The Secrets of Bletchley Park's Codebreaking Computers.* Oxford, England: Oxford University Press.

Copernicus, Nicolaus (1992). *On the Revolutions* (Translation and Commentary by Edward Rosen). Baltimore, MD & London: The Johns Hopkins University Press.

Corsi, P. (Ed.). (1983). *Information Sources in the History of Science and Medicine.* London: Butterworth Scientific.

Cortada, James W. (2004). *The Digital Hand: How Computers Changed the Work of American Manufacturing, Transportation, and Retail Industries.* New York: Oxford University Press.

———— (2004). *The Digital Hand: How Computers Changed the Work of American Public Sector Industries.* New York: Oxford University Press.

———— (2006). *The Digital Hand: How Computers Changed the Work of American Financial, Telecommunications, Media, and Entertainment Industries.* New York: Oxford University Press.

Couldry, Nick, and Anna McCarthy (Eds.). (2004). *Mediaspace: Place, Scale, and Culture in a Media Age.* New York: Routledge.

Crevier, Daniel (1993). *AI: The Tumultuous History of the Search for Artificial Intelligence.* New York: Basic Books.

Crombie, Alistair C. (1992). *Augustine to Galileo* (2 Vols) (M. Iatridou and D. Kourtovik, Trans). Athens: MIET.

Curtin, Jeremiah (1996; 1st ed. 1908). *The Mongols: A History.* Conshohocken, PA: Combined Books.

Curtin, Philip D. (1990). *The rise and fall of the plantation complex: Essays in Atlantic History.* Cambridge: Cambridge University Press, 1990.

Dale, John, and David Kyle (2015). Smart Transitions? Foreign Investment, Disruptive Technology, and Democratic Reform in Myanmar. *Social Research: An International Quarterly* 82(2): pp. 291-326.

Dallal, Ahmad (1999). Science, Medicine, and Technology: The Making of a Scientific Culture. In *The Oxford History of Islam,* edited by John L. Esposito, 155-214. New York: Oxford University Press.

Davies, Martin (2012). *The Universal Computer. The Road from Leibniz to Turing* (2nd ed.). Boca Raton, London, New York: CRC Press.

Davison, George S. (1928). Pittsburgh and the Petroleum Industry. In: Chamber of Commerce of Pittsburgh. *Pittsburgh and the Pitts-*

burgh Spirit: Addresses at the Chamber of Commerce of Pittsburgh, 1927-1928. Pittsburgh, PA: Robert L. Forsythe Company, pp. 85-104.

Day, Christopher, Pam Sammons, Gordon Stobart, Alison Kington, and Qing Gu (2007). *Teachers Matter: Connecting Work, Lives and Effectiveness*. Maidenhead, England: Open University Press.

Day, Lance, and Ian McNeil (Eds.). (1998). *Biographical Dictionary of the History of Technology*. London: Routledge.

DeLancey, Craig (2002). *Passionate Engines: What Emotions Reveal about Mind and Artificial Intelligence*. New York: Oxford University Press.

Deming, David (2010). *Science and Technology in World History*. 3 Vols. Jefferson: McFarland and Company.

Derbyshire, John (2006). *Unknown Quantity: A Real and Imaginary History of Algebra*. Washington D.C.: Joseph Henry Press.

Derry, T. K., and Trevor I. Williams (1961). *A Short History of Technology from the Earliest Times to A.D. 1900*. New York: Oxford University Press.

Di Cosmo, Nicola (2002). *Warfare in Inner Asian History: 500-1800*. Boston: Brill.

Diaper, Dan, and Neville A. Stanton (Eds.). (2004). *The Handbook of Task Analysis for Human-Computer Interaction*. Mahwah, NJ: Lawrence Erlbaum Associates.

Dijksterhuis, Eduard J. (1989). *Die Mechanisierung des Weltbildes*. Berlin, Heidelberg, New York: Springer.

————— (1956). Die Mechanisierung des Weltbildes. *Physikalische Blätter* 11: 481–494.

Diller, A. (1941). The Parallels on the Ptolemaic Maps. *Isis* 33(1): pp. 4-7.

Dimitriadis, N. D. (1983). *Anatomy of rhetoric. The "disagreement" between Plato and Isocrates*. Athens: Michalas.

Dinello, Daniel (2005). *Technophobia!: Science Fiction Visions of Posthuman Technology*. Austin, TX: University of Texas Press.

Disney, Anthony (2010). Prince Henry of Portugal and the Sea Route to India. *Historically Speaking* 11(3): pp. 35-37.

Docherty, Peter, Jan Forslin, and A. B. (Rami) Shani (2002). *Creating Sustainable Work Systems: Emerging Perspectives and Practice*. London: Routledge.

Dodgson, C.L. (2009). *Euclid and His Modern Rivals*. New York: Cambridge University Press.

Donner, Fred M. (1999). Muhammad and the Caliphate: Political History of the Islamic Empire up to the Mongol Conquest. In: *The Oxford History of Islam*, edited by John L. Esposito. New York: Oxford University Press: pp. 1-62.

Dovey, Jon, and Helen W. Kennedy (2006). *Game Cultures: Computer Games as New Media*. Maidenhead, England: Open University Press.

Dubai will precede Turkey in receiving Iranian gas (1997). *Oil,* 130: p. 5.

Duhem, Pierre (1954). *The Aim and Structure of Physical Theory* (Philip P. Wiener, Trans). Princeton, NJ: Princeton University Press.

Dumett, Raymond E. (Ed.). (1999). *Gentlemanly Capitalism and British Imperialism: the New Debate on Empire.* London and New York: Routledge.

Dunn, Ross E. (2012; 1st ed. 1986). *The Adventures of Ibn Battuta.* Berkeley: University of California Press.

Dym, Warren (2005). Scholars and Miners: Dowsing and the Freiberg Mining Academy. *Technology and Culture* 49(4): pp. 833-859.

Earman, J. (1989). *World Enough and Spacetime: Absolute versus Relational Theories of Space and Time.* Cambridge MA: MIT Press.

Edelmayer, Friedrich (2010). Die ,Leyenda negra' und die Zirkulation anti-katholisch-antispanischer Vorurteile. In: Europäische Geschichte Online (EGO), hrsg. vom Institut für Europäische Geschichte (IEG), Mainz 03.12.2010.

Edwards, John (2005). *The Geeks of War: The Secretive Labs and Brilliant Minds behind Tomorrow's Warfare Technologies.* New York: AMACOM.

Egypt: Oil Production is Declining, as the Gas Production Rises, together with the Demand for Gas (1999). *Oil,* 139: p. 6.

Einstein, Albert (1916). Die Grundlage der allgemeinen Relativitätstheorie. *Annalen der Physik* 354(7): pp. 769-822.

Elden, S. and J.W. Crampton (2007). *Space, Knowledge and Power: Foucault and Geography.* Hampshire: Ashgate.

El-hawary, Mohamed E. (2014). The Smart Grid—State-of-the-art and Future Trends. *Electric Power Components and Systems* 42(3-4): pp. 239-250.

Eliade, Mircea (1978). *The Forge and the Crucible.* Chicago, London: University of Chicago Press.

Ercker, Lazarus (1951). *Treatise on Ores and Assaying.* Chicago: University of Chicago Press.

Esteban Piñeiro, Mariano (2002-2003). Las Academias Técnicas en la España del Siglo XVI. *Quaderns d'Història de l'Enginyeria,* Volum V: pp. 10-19.

Euclidis Elementa (1883–1884), edidit et Latine interpretatus est I.L. Heiberg, Lipsiae, in aedibus B.G. Teubneri.

Fara, Patricia (1996). *Sympathetic Transactions: Magnetic Practices, Beliefs, and Symbolism in Eighteenth Century England.* Princeton, NJ: Princeton University Press.

Fabre, André-Julien (2008). Pneumatic Machines in Antiquity (Air as Source of Energy in the Treatise on Pneumatics of Heron of Alexandria). *Analecta Historico Medica* VI (1): pp. 67-70.

Faulkner, W. (1994). Conceptualizing Knowledge Used in Innovation: A Second Look at the Science-Technology Distinction and

Industrial Innovation. *Science, Technology, and Human Values,* 19(4): pp. 425-458.

Feingold, Mordechai, and Navarro Brotons, Víctor (Eds.). (2006). *Universities and Science in the Early Modern Period.* Dordrecht: Springer.

Fernández-Armesto, Felipe (2004). Maritime History and World History. In Daniel Finamore (ed.), *Maritime History as World History.* Gainesville: University Press of Florida: pp. 7-34.

Fernández-Armesto, Felipe (2009). *1492: The Year Our World Began.* New York, NY: Harper Collins.

Feyerabend, Paul (1993). *Against Method* (3rd Ed.). London: Verso.

Feynman, Richard P. (1988). *QED: The Strange Theory of Light and Matter.* Princeton, NJ: Princeton University Press.

Fink, Leon (Ed.). (2011). *Workers across the Americas: The Transnational Turn in Labor History.* New York: Oxford University Press.

Finocchiaro, Maurice A. (1980). *Galileo and the Art of Reasoning: Rhetorical Foundations of Logic and Scientific Method* (Boston Studies in the Philosophy of Science, Vol. 61). Dordrecht and Boston: D. Reidel Publishing Co.

———— (2005). *Retrying Galileo, 1633-1992.* Berkeley, CA: University of California Press.

———— (2010). *Defending Copernicus and Galileo: Critical Reasoning in the Two Affairs* (Boston Studies in the Philosophy of Science, Vol. 280). New York: Springer.

Fisch, M. and S. Schaffer (1991). *William Whewell, a Composite Portrait.* New York: Oxford University Press.

Fischer, I. (1975). Another Look at Eratosthenes' and Posidonius' Determinations of the Earth's Circumference. *Quarterly Journal of the Royal Astronomical Society,* 16: pp. 152-167.

Flach, John, Peter Hancock, Jeff Caird, and Kim Vicente (Eds.). (1995). *Global Perspectives on the Ecology of Human-Machine Systems.* Vol. 1. Hillsdale, NJ: L. Erlbaum Associates.

Fleck, Ludwig (1979). *The Genesis and Development of a Scientific Fact.* Chicago: The University of Chicago Press.

Flynn, Dennis O. and Giráldez, Arturo (1995). Born with a 'Silver Spoon': The Origin of World Trade in 1571. *Journal of World History* 6(2): pp. 201-221.

Foerster, Heinz Von (2014). *The Beginning of Heaven and Earth Has No Name: Seven Days with Second-Order Cybernetics* (Edited by Albert Müller and Karl H. Müller. Translated by Elinor Rooks and Michael Kasenbacher). New York: Fordham University Press.

Fontes da Costa, Palmira, and Henrique Leitão (2009). Portuguese Imperial Science, 1450-1800: A Historiographical Review. In: Daniela Bleichmar et al. (eds.), *Science in the Spanish and Portuguese Empires, 1500-1800.* Stanford, CA: Stanford University Press: pp. 35-53.

Foucault, Michel (1984). Of Other Spaces, Heteroto-
pias. *Architecture, Mouvement, Continuité*, 5: pp. 46-49.

Frank, Philipp (1957). *Philosophy of Science: The Link between Science and Philosophy*. Englewood Cliffs, NJ: Prentice-Hall.

Franks, Kenny A., and Paul F. Lambert (1985). *Early California Oil: A
Photographic History, 1865-1940*. College Station, TX: Texas University Press.

French, Robert M. (2012). Dusting Off the Turing Test. *Science*
336(164): pp. 164-165.

Friedman, M. (1983). *Foundations of Space-Time Theories, Relativistic Physics and Philosophy of Science*. Princeton, NJ.

Fuchsluger, Andreas (2013). *Seefahrt zwischen der Iberischen Halb-
insel und der Neuen Welt in der frühen Neuzeit (1500-1700) als
Träger der Protoglobalisierung*, Diplomarbeit, University of Vienna, Historisch-Kulturwissenschaftliche Fakultät.

Fuller, S., and Collier, J. H. (2004). *Philosophy, Rhetoric, and the End
of Knowledge: A New Beginning for Science and Technology Studies*
(2nd Ed.). Mahwah, NJ: Lawrence Erlbaum Associates.

Furlong, Andy (1992). *Growing Up in a Classless Society? School to
Work Transitions*. Edinburgh: Edinburgh University Press.

Gale, G. (1970). The Physical Theory of Leibniz. *Studia Leibnitiana*
2(2): pp. 114-27.

Galilei, Galileo (1953). *Dialogue on the Great World Systems*. (Trans.
and ed. G. de Santillana). Chicago: University of Chicago Press.

———— (1957). *Discoveries and Opinions of Galileo* (Translated by
Stillman Drake). New York: Doubleday.

———— (1968). *Opere di Galileo Galilei* (Edizione Nazionale edited
by Antonio Favaro. Vol. V). Firenze: Giunti Barbera.

Galison, Peter (1994). The Ontology of the Enemy: Norbert Wiener
and the Cybernetic Vision. *Critical Enquiry*, 21: pp. 229-266.

Gallagher, John, and Ronald Robinson (1953). The Imperialism of
Free Trade. *Economic History Review*, New Series 6(1): pp. 1-15.

Gardner, Howard (1993). *Frames of Mind: The Theory of Multiple
Intelligences* (2nd Ed.). New York: BasicBooks.

Gary, James H., Glenn E. Handwerk, and Mark J. Kaiser (2007).
Petroleum Refining: Technology and Economics (5th Ed.). Boca Raton, FL: CRC Press.

Gaspar, J.A. (2013). From the Portolan Chart to the Latitude Chart.
The silent cartographic revolution. *Comité Français de Cartogra-
phie*, 216: pp. 67-77.

Gaukroger, Stephen (2002). *Descartes' System of Natural Philosophy*.
Cambridge, England: Cambridge University Press.

Gaynor, Jennifer L. (2013). Ages of Sail, Ocean Basins, and Southeast
Asia. *Journal of World History* 24(2): pp. 309-333.

Geraci, Robert M. (2010). *Apocalyptic AI: Visions of Heaven in Robotics, Artificial Intelligence, and Virtual Reality*. New York: Oxford
University Press.

Gerali, Francesco (2011). Scientific Maturation and Production Modernization; Notes on the Italian Oil Industry in the XIXTH Century. *Oil-Industry History* 12(1): 89-108.

Gerber, Rod, and Colin Lankshear (2000). *Training for a Smart Workforce.* London: Routledge.

Germany: Decreasing demand for Gas Oil (1997). *Oil,* 130: p. 4.

Gesner, Abraham (1861). *A Practical Treatise on Coal, Petroleum, and other Distilled Oils.* New York: Baillière Brothers.

Gilbert, William (1893). *On the Loadstone and Magnetic Bodies, and on the Great Magnet the Earth.* Translated by P. Fleury Mottelay. London: Bernard Quaritch.

Gille, Bertrand (1947). *Les origines de la grande metallurgie en France.* Paris: Éditions Domat Montchrestien.

Gillispie, Charles C. (1990). *The Edge of Objectivity: An Essay in the History of Scientific Ideas* (10th Ed.). Princeton, NJ: Princeton University Press.

Ginev, Dimitri (2016). Hermeneutic Perspectives on Science in Fleck's Work and Hermeneutic Critic of Constructivist Epistemology. *Perspectives on Science* 14(2): pp. 228-253.

Gingerich, Owen (2002). Kepler Then and Now. *Perspectives on Science* 10(2): pp. 228-240.

———— (2004). *The Book Nobody Read: Chasing the Revolutions of Nicolaus Copernicus.* New York: Walker & Company.

Gladney, Dru C. (1999). Central Asia and China: Transnationalization, Islamization and Ethnicization. In *The Oxford History of Islam,* edited by John L. Esposito, 433-474. New York: Oxford University Press.

Glover, John George, and William Bouck Cornell (Eds.). (1941). *The Development of American Industries, Their Economic Significance.* New York: Prentice-Hall, 1941.

Goenner, Hubert F.M. (2004). On the History of Unified Field Theories. *Living Reviews in Relativity* 7(2): pp. 1-153.

Gold, Thomas (1992). The deep, hot biosphere. *Proceedings of the National Academy of the Sciences* 89: pp. 6045-6049.

———— (1999). *The deep hot biosphere: The Myth of Fossil Fuels.* New York: Springer.

Goldstein, B.R. and Bowen, A.C. (1983). A New View of Early Greek Astronomy. *Isis,* 74 (3): pp. 330-340.

Goldstein, Catherine, Norbert Schappacher, and Joachim Schwermer (2007). *The Shaping of Arithmetic after C.F. Gauss's Disquisitiones Arithmeticae.* Berlin, Heidelberg: Springer.

Goldstine, Herman H. (1993). *The Computer from Pascal to von Neumann* (5th Ed.). Princeton, NJ: Princeton University Press.

Goodman, David (2009). Science, Medicine, and Technology in Colonial Spanish America: New Interpretations, New Approaches. In: Daniela Bleichmar et al. (ed.), *Science in the Spanish and Por-*

tuguese Empires, 1500-1800. Stanford, CA: Stanford University Press: pp. 9-34.

Goodrich, L. Carrington, and Fêng Chia-Shêng (1946). The Early Development of Firearms in China. *Isis* 36(2): pp. 114-123.

Goodwin, Richard M. (1990). *Chaotic Economic Dynamics.* Oxford, England: Clarendon.

Gough, Barry M. (1992). *The Northwest Coast: British Navigation, Trade, and Discoveries to 1812.* Vancouver, BC: University of British Columbia Press.

Gräßner, C.A. (2011). Wissensräume, Raumwissen und Wissensordnungen. Historisch-kulturwissenschaftliche Forschungen zum Korrelat Raum — Wissen, *eTopoi,* 1: pp. 105–113.

Gray, Asa (1880). Biographical Memorial, in behalf of the Board of Regents. In: Smithsonian Institution (Ed. *A Memorial of Joseph Henry).* Washington D.C.: Government Printing Office: pp. 53-73.

Green, Bert F. (1963). *Digital Computers in Research: An Introduction for Behavioral and Social Scientists.* New York: McGraw-Hill.

Greer, Margaret R., Walter D. Mignolo, and Maureen Quilligan (Eds.). (2007). *Rereading the Black Legend: The Discourses of Religious and Racial Difference in the Renaissance Empires.* Chicago: The University of Chicago Press.

Grossman, Wendy M. (2001). *From Anarchy to Power: The Net Comes of Age.* New York: New York University Press.

Guastello, Stephen J. (1995). *Chaos, Catastrophe, and Human Affairs: Applications of Nonlinear Dynamics to Work, Organizations, and Social Evolution.* Mahwah, NJ: Lawrence Erlbaum Associates.

Gulbekian, E. (1987). The Origin and Value of the Stadion Unit used by Eratosthenes in the Third Century BC. *Archive for History and Exact Sciences,* 37 (4): pp. 359-363.

Gunkel, David J. (2001). *Hacking Cyberspace.* Boulder, CO: Westview Press.

Gupta, A. and J. Ferguson (1992). Beyond 'Culture': Space, Identity, and the Politics of Difference. *Cultural Anthropology,* 7(1): pp. 6-23.

Hacking, Ian (1995). *The Emergence of Probability. A Philosophical Study of Early Ideas about Probability, Induction and Statistical Inference* (6th Ed.). Cambridge, New York, Melbourne: Cambridge University Press.

Hadzsits, G. D. (1963). *Lucretius and his Influence.* New York: Cooper Square Publishers.

Haggett, N. (Ed.). (1999). *Space from Zeno to Einstein: classic readings with a contemporary commentary by Nick Haggett.* Cambridge, MA: MIT Press.

Hahn, Martin, and Scott C. Stoness (Eds.). (1999). *Proceedings of the Twenty First Annual Conference of the Cognitive Science Society.* Mahwah, NJ: Lawrence Erlbaum Associates.

Hailperin, Theodore (1976). *Boole's Logic and Probability. A Critical Exposition from the Standpoint of Contemporary Algebra, Logic and Probability Theory.* Amsterdam, New York, Oxford: North-Holland Publishing Company.

Hall, E.H. (1997). On a new Action of the Magnet on Electric Currents. In *Science in the Making: Scientific Development as Chronicled by Historic Papers in the Philosophical Magazine, with Commentaries and Illustrations,* vol. 2, edited by E.A. Davis. London: Taylor and Francis: pp. 141-146.

Hallevy, Gabriel (2013). *When Robots Kill: Artificial Intelligence under the Criminal Law.* Boston: Northeastern University Press.

Harman, Peter M. (1994). *Energy, Force and Matter: The Conceptual Development of Nineteenth-Century Physics* (Translated by Tasos Tsiantoulas). Heraklion: Crete University Press.

Harrison, Andrew, Paul Wheeler, and Carolyn Whitehead (Eds.). (2003). *The Distributed Workplace: Sustainable Work Environments.* New York: Routledge.

Hartz, G. A. and J. A. Cover (1988). Space and Time in the Leibnizian metaphysics. *Nous* 22(4): pp. 493-519.

Haugeland, John (1989). *Artificial Intelligence, the Very Idea.* Athens: Katoptro.

Hawke, Jason G. (2008). Number and Numeracy in Early Greek Literature. *Syllecta Classica* 19: pp. 1-76.

Hawking, S.W. (1988). *A Brief History of Time. From the Bing Bang to Black Holes.* London: Bantam.

————— (2003). *On the Shoulders of Giants. The Great Works of Physics and Astronomy.* London: Penguin

Hawking, S.W. and Ellis, G.F.R. (1973). *The Large Scale Structure of Space-Time.* Cambridge: University Press.

Hawley, John F., and Katherine A. Holcomb (1998). *Foundations of Modern Cosmology.* New York: Oxford University Press.

Headrick, Daniel R. (2009). *Technology: A World History.* New York: Oxford University Press.

Heath, Thomas (1921). *A History of Greek Mathematics.* Oxford: Clarendon Press, 1921.

Heilbron, John L. (1990). Introductory Essay. In: Frängsmyr, Tore, John L. Heilbron and Robin E. Rider (Eds). *The Quantifying Spirit in the 18th Century.* Berkeley: University of California Press.

Heller, Henry (1996). *Labour, Science and Technology in France, 1500-1620.* New York: Cambridge University Press.

Hempel, C. G. (2001). *The Philosophy of Carl G. Hempel: Studies in Science, Explanation, and Rationality* (J. H. Fetzer, Ed.). New York: Oxford University Press.

Hennig, Richard (1936-39; 1944-1956). *Terrae incognitae: eine Zusammenstellung und kritische Bewertung der wichtigsten vorcolumbischen Entdeckungsreisen an Hand der darüber vorliegenden Originalberichte.* 4 Bde. Leiden: Brill.

Hentschel, Klaus (2003). Der Vergleich als Brücke zwischen Wissenschaftsgeschichte und Wissenschaftstheorie. *Journal for General Philosophy of Science* 34: pp. 251–275.

Heppenheimer, T. A. (2001). *A Brief History of Flight: From Balloons to Mach 3 and Beyond.* New York: Wiley.

Hicks, Marie (2010). Repurposing Turing's 'Human Brake'. *IEEE Annals of the History of Computing* 32(4): pp. 108, 106.

Hill, Donald R. (1998). *Studies in Medieval Islamic Technology: From Philo to al-Jazari, from Alexandria to Diyar Bakr.* Aldershot, Hants.

Hodges, Andrew (2004). *Alan Turing. The Enigma.* Athens: Travlos.

———— (2012). Beyond Turing's Machines. *Science* 336(163): pp. 163-164.

Hodgson, Marshall G.S. (1974). *The Venture of Islam. Conscience and History in a World Civilization.* Vol. 3: *The Gunpowder Empires and Modern Times.* Chicago: The University of Chicago Press.

Hodgson, Peter (2005). Galileo the Theologian. *Logos: A Journal of Catholic Thought and Culture* 8(1): pp. 28-51.

Hooykaas, Reijer (1972). *Religion and the Rise of Modern Science.* Edinburgh: Scottish Academic Press.

———— (1983). *The Portuguese Discoveries and the Rise of Modern Science.* In: Reijer Hooykaas, *Selected Studies in History of Science.* Coimbra Universidade.

———— (1999). *Fact, Faith and Fiction in the Development of Science: The Gifford Lectures Given in the University of St Andrews 1976.* Springer.

Horvitz, Leslie Alan (2002). *Eureka! Stories of Scientific Discovery.* New York: Wiley.

Hoschka, Peter (Ed.) (1996). *Computers as Assistants: A New Generation of Support Systems.* Mahwah, NJ: Lawrence Erlbaum Associates.

Howard, Nicole (2008). Marketing Longitude: Clocks, Kings, Courtiers, and Christian Huygens. *Book History* 11: pp. 59-88.

Huizinga, Johan (1949). *Homo Ludens. A Study of the Play-Element in Culture.* London, Boston and Henley: Routledge and Kegan Paul.

Hutchin, E. (1993). Learning to navigate. In *Understanding Practice. Perspectives on Activity and Context.* (Edited by S. Chaiklin and J. Lave). London: Cambridge University Press.

Iran: Increase of the preferences for the oil pipeline crossing Kazakhstan through the territory of the country (1999). *Oil,* 139: pp. 6-7.

Iran: Production problems decrease exports of light crude oil (1997). *Oil,* 130: pp. 2-3.

Jacko, Julie A., and Andrew Sears (Eds.). (2003). *The Human-Computer Interaction Handbook: Fundamentals, Evolving Tech-*

nologies, and Emerging Applications. Mahwah, NJ: Lawrence Erlbaum Associates.

Jagacinski, Richard J., and John M. Flach (2003). *Control Theory for Humans: Quantitative Approaches to Modeling Performance.* Mahwah, NJ: Lawrence Erlbaum Associates.

Jammer, M. (1993). *Concepts of Space. The History of Theories of Space in Physics.* New York: Dover.

Jiang, Xiaoyuan (2015). Astronomy. In: *A History of Chinese Science and Technology* (Ed. Yongxiang Lu; Trans. Chuijun Qian, Qingping Hu, Xiaodi Li, Yao Wang and Liang Zhao), Vol. I. Berlin, Heidelberg: Springer: pp. 41-120.

Japan: The nuclear accident may mean more expanded natural gas for the country (1999). *Oil,* 139: pp. 7-8.

Johnson, Christine R. (2006). Renaissance German Cosmographers and the Naming of America. *Past and Present* 191: pp. 3-43.

Johnston, John (2002). A Future for Autonomous Agents: Machinic Merkwelten and Artificial Evolution. *Configurations* 10(3): pp. 473-516.

Jolley, N. (Ed.). (1995). *The Cambridge Companion to Leibniz.* Cambridge University Press.

Jones, David Martin (2001). *The Image of China in Western Social and Political Thought.* New York: Palgrave.

Kaiser, Paul J. (1996). *Culture, Transnationalism, and Civil Society: Aga Khan Social Service Initiatives in Tanzania.* Westport, CT: Praeger.

Katz, Victor J. and Karen Hunger Parshall (2014). *Taming the Unknown: History of Algebra from Antiquity to the Early Twentieth Century.* Princeton, NJ: Princeton University Press.

Kellenbenz (Ed.). (1974). *Schwerpunkte der Eisengewinnung und Eisenverarbeitung in Europa: 1500-1650.* Cologne.

Kennedy, G. (2004). *History of Classical Rhetoric.* Athens: Papadimas.

Kennefick, Daniel (2009). Testing relativity from the 1919 eclipse - a question of bias. *Physics Today* 62(3): pp. 37–42.

Kenney, Amanda (2013). Encoding Authority: Negotiating the Uses of Khipu in Colonial Peru. *Traversea* 3: pp. 4-19.

Kesrouani, Pamela. How do you Build a Smart City, *Wamda* (17 August 2015).
http://www.wamda.com/memakersge/2015/08/how-do-you-build-a-smart-city

Khan, Iqtidar Alam (1996). The Coming of Gunpowder to the Islamic World and India: Spotlight on the Role of the Mongols. *Journal of Asian History* 30: pp. 27–45.

Kimmel, Jean, and Emily P. Hoffman (Eds.) (2002). *The Economics of Work and Family.* Kalamazoo, MI: W.E. Upjohn Institute for Employment Research.

Kindleberger, Charles P. (1984). *A Financial History of Western Europe*. London: George Allen and Unwin.

Kipiniak, Walerian (1961). *Dynamic Optimization and Control: A Variational Approach*. Cambridge: M.I.T. Press.

Kirlik, Alex (Ed.). (2006). *Adaptive Perspectives on Human-Technology Interaction: Methods and Models for Cognitive Engineering and Human-Computer Interaction*. New York: Oxford University Press.

Klee, R. (1997). *Introduction to the Philosophy of Science: Cutting Nature at Its Seams*. New York: Oxford University Press.

Klein, K.J. and J.S. Sorra (1996). The Challenge of Innovation Implementation. *The Academy of Management Review*, 21(4): pp. 1055-1080.

Kline, Ronald (2000). *Consumers in the Country, Technology and Social Change in Rural America*. Baltimore and London: he John Hopkins University Press.

Knorr, Wilbur R. (1986). *The Ancient Tradition of Geometric Problems*. New York: Dover.

———— (1993). Arithmetike stoicheiosis: On Diophantus and Hero of Alexandria. *Historia Mathematica* 20: pp. 180-192.

Koyré, A. (1939). *Études galiléennes*. 3 Vols. Paris: Hermann.

———— (1957). *From the Closed World to the Infinite Universe*. Baltimore: Johns Hopkins University Press.

———— (1968). *Newtonian Studies*. University of Chicago Press.

Kraut, Robert, Malcolm Brynin, and Sara Kiesler (Eds.). (2006). Computers, Phones, and the Internet: Domesticating Information Technology. In: *Computers, Phones, and the Internet: Domesticating Information Technology* (Edited by Robert Kraut, Malcolm Brynin, and Sara Kiesler), Iii-Iv. New York: Oxford University Press.

Kropf, Marianna (2005). *Rituelle Traditionen der Planetengottheiten (Navagraha) im Kathmandutal Strukturen – Praktiken – Weltbilder*. Inaugural Dissertation, Universität Heidelberg.

Kuhn, Thomas S. (1977). *The Essential Tension: Selected Studies in Scientific Tradition and Change*. Chicago and London: The University of Chicago Press.

———— (1981). *Die kopernikanische Revolution* (übersetzt von Helmut Kühnelt). Wiesbaden: Springer.

———— (1996). *The Structure of Scientific Revolutions* (3rd Ed.). Chicago and London: The University of Chicago Press.

Kulp, C. B. (1992). *The End of Epistemology: Dewey and His Current Allies on the Spectator Theory of Knowledge*. Westport, CT: Greenwood Press.

Ladyman, J. (2002). *Understanding Philosophy of Science*. London: Routledge.

Lajoie, Susanne P. (Ed.), (2000). *Computers as Cognitive Tools: No More Walls: Theory Change, Paradigm Shifts, and Their Influence on the Use of Computers for Instructional Purposes* (2nd ed. Vol. 2). Mahwah, NJ: Lawrence Erlbaum Associates.

Lakatos, Imre (1968). Criticism and Methodology of Scientific Research Programmes. *Proceedings of the Aristotelian Society* (London: Blackwell Publishing), 69: 149-186.

Latour, Bruno (1998). From the World of Science to the World of Research? *Science* 280(5361).

———— (2005). *Reassembling the Social: An Introduction to Actor-Network-Theory.* Oxford: Oxford University Press.

Laudan, Larry (1990). The History of Science and the Philosophy of Science. In R.C. Olby, G.N. Cantor, J.R.R. Christie, M.J.S. Hodge. *Companion to the History of Modern Science.* London: Routledge.

Launius, Roger D., and Howard E. McCurdy (2008). *Robots in Space: Technology, Evolution, and Interplanetary Travel.* Baltimore: Johns Hopkins University Press.

Lazos, C.D. (1995). *Archimedes. The Ingenious Engineer.* Athens: Aiolos.

Le Roy, G. (Ed.). (1966). *Leibniz, Discours de Metaphysique et Correspondence avec Arnauld.* Paris: Librairie Philosophique J. Vrin.

Leclerc I. (1973). *Leibniz and the Analysis of Matter and Motion.* In: Leclerc I. (Ed.), *The Philosophy of Leibniz and the Modern World.* Nashville: Vanderbilt University Press: 114-32.

Leibniz, G. W. (1966). *Hauptschriften zur Grundlengung der Philosophie.* Hamburg: Felix Meiner Verlag.

———— (1969). *Die philosophischen Schriften* (Bund 7). Hildesheim: Georg Olms.

Lennox, J. G. (2011). Aristotle on Norms of Inquiry. *HOPOS: The Journal of the International Society for the History of Philosophy of Science,* 1(1): pp. 23-46.

Lerski, George J. (1996). *Historical Dictionary of Poland, 966-1945.* Westport, CT: Greenwood Press.

Lesky, A. (1998). *History of Ancient Greek Literature.* Thessaloniki: Kyriakidis Brothers.

Light, Jennifer S. (1999). When Computers were Women. *Technology and Culture,* 40: pp. 455-483.

Lin, Tzung-De (2015). Theater as a Site for Technology Demonstration and Knowledge Production: Theatrical Robots in Japan and Taiwan, *East Asian Science, Technology and Society: an International Journal* 9(2): pp. 187-211.

Lindberg, David C. (2007). *The Beginnings of Western Science: The European Scientific Tradition in Philosophical, Religious, and Institutional Context, Prehistory to A.D. 1450* (2nd Ed.). Chicago: The University of Chicago Press.

Ling, Wang (1947). On the Invention and use of Gunpowder and Firearms in China. *Isis* 37 (3/4): pp. 160-178.

Lipartito, Kenneth (2003). Picturephone and the Information Age. In: *Technology and Culture* 44: pp. 50-81.

Liu, Dun (2015). Vertical and Horizontal Beginnings. In: *A History of Chinese Science and Technology* (ed. Yongxiang Lu, trans. Chuijun Qian, Qingping Hu, Xiaodi Li, Yao Wang and Liang Zhao), Vol. I. Berlin, Heidelberg: Springer: pp. 1-40.

Liu, Haiming (2005). *The Transnational History of a Chinese Family: Immigrant Letters, Family Business, and Reverse Migration.* New Brunswick, NJ: Rutgers University Press.

Livingston, David (2013). *Putting Science in its Place: Geographies of Scientific Knowledge.* Chicago: University of Chicago Press.

Livingstone, David N., D. G. Hart, and Mark A. Noll (Eds.). (1999). *Evangelicals and Science in Historical Perspective.* New York: Oxford University Press, 1999.

Long, Pamela O. (2003). Of Mining, Smelting, and Printing: Agricola's De re Metallica. *Technology and Culture* 44(1): pp. 97-101.

Lu, Yongxiang (ed.) (2015). *A History of Chinese Science and Technology* (4 Vols.). Translated by Chuijun Qian, Qingping Hu, Xiaodi Li, Yao Wang and Liang Zhao. Berlin, Heidelberg: Springer.

Lucas, Adam R. (2005). Industrial Milling in the Ancient and Medieval Worlds. A Survey of the Evidence for an Industrial Revolution in Medieval Europe. *Technology and Culture* 46(1): pp. 1-30.

Lucier, Paul (2008). *Scientists and Swindlers: Consulting on Coal and Oil in America, 1820–1890.* Baltimore, MA: The John Hopkins University Press.

McCormick, John (2012). *Nine Algorithms That Changed the Future: The Ingenious Ideas That Drive Today's Computers.* Princeton, NJ: Princeton University Press.

Mach, Ernst (1883; 7th ed., 1912). *Die Mechanik in ihrer Entwicklung historisch-kritisch dargestellt.* Leipzig.

Machado, Pedro (2005). *Gujarati Indian Merchant Networks in Mozambique, 1777–c. 1830.* PhD diss., University of London.

Mackenzie, Donald, and Judy Wajcman (Eds.). *The Social Shaping of Technology* (2nd edition). Buckingham: Open University Press.

Maerivoet, Sven (2001). *Leidt het Standaardmodel in de elementaire deeltjesfysica tot een quantum veldentheorie van Alles?* Universitaire Instelling Antwerpen.

Maimonides, Moses (n.d.). *The Guide for the Perplexed. Grand Rapids, MI: Christian Classics Ethereal Library. Chapter IX. On the Number of the Heavenly Spheres.* Grand Rapids, MI: Christian Classics Ethereal Library, www.ccel.org

Mair, Victor H. (2001). *The Columbia History of Chinese Literature.* New York: Columbia University Press.

Mansell, Robin, and Uta When (1998). *Knowledge Societies: Information Technology for Sustainable Development.* Oxford: Oxford University Press.

Marboe, Alexander (2009). Zur Einführung: Schiffsbau und Nautik im vorneuzeitlichen Europa. In: Alexander Marboe and Andreas

Obenaus (Hrsg.), *Seefahrt und die frühe europäische Expansion.* Wien: Mandelbaum: pp. 11-35.

Markovits, Claude (2000). *The Global World of Indian Merchants, 1750-1947: Traders of Sind from Bukhara to Panama.* Cambridge: Cambridge University Press.

Marques, Alfredo P. (1995). The discovery of the Azores and its first repercussions in cartography. *ARQUIPÉLAGO, História,* 2ª série, vol. 1, nº 2: 7-15. http://hdl.handle.net/10400.3/485

Marrou, H. I. (1961). *History of education in antiquity* (Th. Foteino-poulos, Trans.). Athens.

Martin, G. (1966). *Leibniz, Logique et Métaphysique* (traduit par M. Regnier). Paris: Beauchesne.

Marx, Karl (2011; first published 1867). *Capital. A Critique of Political Economy.* Vol I. (Translated by Samuel Moore and Edward Aveling). Mineola, New York: Dover.

Mason, Stephen F. (1962). *A History of the Sciences.* New York: Collier Books.

Mather, David (2006). Early Calculating Engines and Historical Computer Simulations. *Leonardo,* 39(3): pp. 237-243.

Matthews, Gerald, Moshe Zeidner, and Richard D. Roberts (Eds.). (2007). *The Science of Emotional Intelligence: Knowns and Unknowns.* New York: Oxford University Press.

Mann, Alfred M. (2009). Some Petroleum Pioneers of Pittsburgh. *Oil Industry History* 10(1): pp. 49-68.

Mayr, Otto (1971). Adam Smith and the Concept of the Feedback System. *Technology and Culture* 12 (1): pp. 1-22.

———— (1970). The Origins of Feedback Control. *Scientific American* 223(4), October: pp. 110–118.

McClellan, James E. III, and Dorn, Harold (2006). *Science and Technology in World History. An Introduction* (2nd Ed.). Baltimore: John Hopkins University Press.

McGinnis, J. (Ed.). (2004). *Interpreting Avicenna: Science and Philosophy in Medieval Islam: Proceedings of the Second Conference of the Avicenna Study Group* (Islamic Philosophy, Theology, and Science). Boston: Brill.

McLaughlin, Glenn E. (1937). The Economic Significance of Oil and Gas. In: The Subcommittee on Technology to the National Resources Committee. *Technological Trends and National Policy: Including the Social Implications of New Inventions. June 1937.* Washington, DC: U.S. Government Printing Office: 123-127.

McMullin, Ernan (1998). Galileo on science and Scripture. In: Peter K. Machamer. *The Cambridge Companion to Galileo.* Cambridge; New York: Cambridge University Press: pp. 271-347.

McNeill, William H. (1982). *The Pursuit of Power. Technology, Armed Force, and Society since A.D. 1000.* Chicago: The University of Chicago Press.

McRae, R. (1979). Time and the Monad. *Nature and System* 1(2): pp. 103-09.

Meyer, Bertrand (2009). *Touch of Class: Learning to Program Well with Objects and Contracts.* Dordrecht, Heidelberg: Springer.

Michael, Mike (2006). *Technoscience and Everyday Life: The Complex Simplicities of the Mundane.* Maidenhead, England: Open University Press.

Miller, Gordon L. (1990). Charles Babbage and the design of Intelligence Computers and Society in 19th-Centuty England. *Bulletin of Science and Technology Society* 10: pp. 68-75.

Missiaen, Tine, Ine Demerre, and Valentine Verrijken (2012). Integrated assessment of the buried wreck site of the Dutch East Indiaman 't Vliegent Hart. *Relicta* 9: pp. 191-208.

MIT, SENSEable City Laboratory, *Trash Track* (2009). http://senseable.mit.edu/trashtrack/

Montague, Gilbert Holland (1903). *The Rise and Progress of the Standard Oil Company.* New York: Harper and Brothers.

Morrow, Glenn R. (1992). *Proclus: A Commentary on the First Book of Euclid's Elements.* Princeton, NJ: Princeton University Press.

Morse, Suzanne W. (2004). *Smart Communities: How Citizens and Local Leaders Can Use Strategic Thinking to Build a Brighter Future.* San Francisco: Jossey-Bass.

Morus, Iwan Rhys (Ed.). (2002). *Bodies/Machines.* New York: Berg.

Moser, P. K. (Ed.). (2002). *The Oxford Handbook of Epistemology.* New York: Oxford University Press.

Moss, Jean Dietz (1983). Galileo's Letter to Christina: Some Rhetorical Considerations. *Renaissance Quarterly* 36(4): pp. 547-576.

Mumford, Lewis (1966). Technics and the Nature of Man. *Technology and Culture* 7(3): pp. 303–317.

Mungello, David E. (1989). *Curious Land: Jesuit Accommodation and the Origins of Sinology.* Honolulu: University of Hawaii Press.

Nason, J. W. (1946). Leibniz's Attack on the Cartesian Doctrine. *Journal of the History of Ideas* 7: pp. 447-83

National Academy of Science (1913). *A History of the First Half-Century of the National Academy of Sciences, 1863-1913.* Edited by Frederick W. True. Washington, DC: N.p.

Navarro Brotons, Víctor (1992). Astronomía y Cosmología en la España del Siglo XVI. Seminario «Orotava» de Historia de la Ciencia - Año XI-XII. Actes de les II trobades d'història de la ciència i de la tècnica: (Peníscola, 5-8 desembre 1992) / coord. por Víctor Navarro Brotóns, 1994: pp. 39-52.

———— (2000). Astronomía y cosmografía entre 1561 y 1625. Aspectos de la actividad de los matemáticos y cosmógrafos españoles y portugueses. Cronos: Cuadernos valencianos de historia de la medicina y de la ciencia, ISSN 1139-711X, Vol. 3, Nº. 2, Pp. 349-380.

Needham, Joseph (1986). *Science and Civilisation in China*, vol. 5, *Chemistry and Chemical Technology*, Part 7: *Military Technology; The Gunpowder Epic* (with the collaboration of Ho Peng Yoke, Lu Gwei-Djen and Wang Ling). Cambridge, England: Cambridge University Press.

———— (1954-98). *Science and Civilisation in China*. 7 vols. in 15 parts. Cambridge: Cambridge University Press.

Nef, John U. (1932). *The Rise of the British Coal Industry*. London: Routledge and Kegan Paul.

Netz, Reviel (1998). Deuteronomic Texts: Late Antiquity and the History of Mathematics. *Revue d'histoire des mathématiques* 4: pp. 261-288.

Neugebauer, Otto (1975). *A History of Ancient Mathematical Astronomy*. Heidelberg: Springer.

Neurath, Otto (1973). On the Foundations of the History of Optics. In: *Empiricism and Sociology*, Trans. Paul Foulkes and Marie Neurath. Dordrecht: D. Reidel, pp. 101–12. ["Prinzipielles zur Geschichte der Optik." *Archiv für die Geschichte der Naturwissenschaften und der Technik* 5 (1915): pp. 371–89.]

———— (1983). Classification of Systems of Hypotheses. In *Philosophical Papers*, 1913–1946. Trans. Robert s. Cohen. Dordrecht: D. Reidel. [Zur Klassifikation von Hypothesensystemen (mit besonderer Berücksichtigung der Optik). *Jahrbuch der Philosophischen Gesellschaft an der Universität zu Wien* (1915): pp. 38–63]

Newton, Isaac (1968; 1729). *The Mathematical Principles of Natural Philosophy. Newton's Principles of Natural Philosophy*. London: Dawsons of Pall Mall.

Nigeria: Delays in implementing petrochemical and energy programs because of the monetary and economic crisis (1997). *Oil*, 130: p. 5.

Norman, Donald (1999). *The Invisible Computer*. Cambridge, Mass.: MIT Press.

Norman, Donald A., and Stephen W. Draper (Eds.). (1986). *User Centered System Design: New Perspectives on Human-Computer Interaction*. Hillsdale, NJ: Lawrence Erlbaum Associates.

Northrop F. S. G. (1946). Leibniz's Theory of Space. *Journal of the History of Ideas* 7: pp. 422-46.

Nowell, C.E. (1954). *The Great Discoveries and the First Colonial Empires*. Ithaca and London: Cornell University Press.

Nye, David E. (2006). *Technology Matters, Questions to live with*. Cambridge, Massachusetts: The MIT Press.

O'Leary, Jaime (1997). Basque Whaling in Red Bay, Labrador. *Exploration and Settlement*. http://www.heritage.nf.ca/exploration/basque.html

O'Neill, O. (1996). *Towards Justice and Virtue*. Cambridge University Press.

Ohlman, Herbert (1990). Information: Timekeeping, Computing, Telecommunications and Audiovisual Technologies. In: Ian McNeil, *An Encyclopaedia of the History of Technology* (London, New York: Routledge): p. 703.

Olby, R. C., G. N. Cantor, J. R. Christie, and M. J. Hodge (Eds.). (1990). *Companion to the History of Modern Science.* London: Routledge.

Olien, Diana Davids, and Roger M. Olien (2002). *Oil in Texas: The Gusher Age, 1895–1945.* Austin: University of Texas Press.

Olien, Roger M., and Diana Davids Olien (2000). *Oil and Ideology: The Cultural Creation of the American Petroleum Industry.* Chapel Hill and London: The University of North Carolina Press.

Oliver, John W. (1956). *History of American Technology.* New York: Ronald Press.

Orser, Charles E. Jr. (2002). Padre Island shipwrecks, Texas, USA. n: Charles E. Jr. Orser (Ed.), *Encyclopedia of Historical Archaeology.* London and New York: Routledge: p. 412.

Ouma, Stefan (2012). „Markets in the Making“: Zur Ethnographie alltäglicher Marktkonstruktionen in organisationalen Settings. *Geographica Helvetica* 67: pp. 203-211.

Padrón, Ricardo (2004). *The Spacious Word: Cartography, Literature, and Empire in Early Modern Spain.* Chicago: The University of Chicago Press.

Papadimitriou, Christos, Sanjoy Dasgupta, und Umesh Vazirani (2010). *Algorithms.* New York: McGraw Hill.

Papert, Seymour (1993). *Mindstorms: Children, Computers, and Powerful Ideas* (2nd Ed.). New York: Basic Books.

Parasuraman, Raja, and Mustapha Mouloua (Eds.) (1996). *Automation and Human Performance: Theory and Applications.* Mahwah, NJ: Lawrence Erlbaum Associates.

Parry, John H. (1961). *The establishment of the European Hegemony, 1415-1715: Trade and exploration in the Age of the Renaissance.* New York and Evanston: Harper and Row.

——— (1963). *The Age of Reconnaissance.* Cleveland, OH: World Pub. Co.

——— (1981). *The Discovery of the Sea.* Berkeley, Los Angeles, New York: University of California Press.

Partington, J. R. (1960). *A History of Greek Fire and Gunpowder.* Cambridge.

Paterson, Michael (2011). *Voices of the Code Breakers: Personal Accounts of the Secret Heroes of World War II.* Cincinnati, OH: David and Charles.

Payr, Sabine, and Robert Trappl (Eds.). (2004). *Agent Culture: Human-Agent Interaction in a Multicultural World.* Mahwah, NJ: Lawrence Erlbaum.

Pearson, K. (1957). *The Grammar of Science.* New York: Meridian Books.

Perkins, Franklin (2004). *Leibniz and China. A Commerce of Light.* Cambridge: Cambridge University Press.

Peters, Tom (1997). *Circle of Innovation: You Can't Shrink Your Way to Greatness.* London: Hodder and Stoughton.

Pielou, E.C. (2001). *The Energy of Nature.* Chicago: University of Chicago Press, 2001.

Pinsky, Michael (2003). *Future Present: Ethics And/As Science Fiction.* Madison, NJ: Fairleigh Dickinson University Press.

Pirages, Dennis C. (Ed.). (1996). *Building Sustainable Societies: A Blueprint for a Post-Industrial World.* Armonk, NY: M. E. Sharpe.

Plato (1939). *Gorgias.* Athens: Zacharopoulos.

———— (1956). *Euthedemus.* Athens: Zacharopoulos.

———— (1993). *Phaedrus.* Athens: Cactus.

———— (1993). *Sophist.* Athens: Cactus.

———— (2004). *Symposium.* Thessaloniki: Zitros.

———— (n.d.). *Laches - Menon – Parmenides.* Athens: Zacharopoulos.

Poincaré, Henri (1952). *Science and Hypothesis.* Trans. William John Greenstreet. New York: Dover. [*La science et l'hypothèse.* Paris: Flammarion, 1902.]

———— (1958). *The Value of Science.* Trans. G. B. Halsted. New York: Dover.

Popper, Karl (2010). *The Logic of Scientific Discovery.* New York: Routledge.

Portuondo, María M. (2009a). Cosmography at the Casa, Consejo, and Corte during the Century of Discovery. In: Daniela Bleichmar et al. (ed.), *Science in the Spanish and Portuguese Empires, 1500-1800.* Stanford, CA: Stanford University Press: pp. 57-77.

———— (2009b). *Secret science: Spanish cosmography and the new world.* Chicago: The University of Chicago Press, 2009.

Potter, V. G. (1996). *Peirce's Philosophical Perspectives* (V. M. Colapietro, Ed.). New York: Fordham University Press.

Powell, Barry B. (2012). *Writing: Theory and History of the Technology of Civilization.* Chichester, West Sussex, UK: Wiley-Blackwell.

Prestage, Edgar (1933). *The Portuguese Pioneers.* London: A. and C. Black.

Psillos, S. (2007). *Philosophy of Science A-Z.* Edinburgh: Edinburgh University Press.

Ptak, Roderich (2007). *Die Maritime Seidenstraße: Küstenräume, Seefahrt und Handel in vorkolonialer Zeit.* München: C.H. Beck.

Pyenson, L., and Verbruggen, C. (2011). Elements of the Modernist Creed in Henri Pirenne and George Sarton. *History of Science, 49*(165): p. 377.

Quataert, D. (1977). Limited Revolution: The Impact of the Anatolian Railway on Turkish Transportation and the Provisioning of

Istanbul, 1890-1908. *The Business History Review*, 51(2): pp. 139-160.

Rabinowitz, M. (2007). Deterrents to a Theory of Quantum Gravity. *International Journal of Theoretical Physics* 46 (5): pp. 1403-1415.

Raina, Dhruv (2003). Betwixt Jesuit and Enlightment Historiography: Jean-Sylvain Bailly's History of Indian Astronomy. *Revue d'histoire des mathématiques* 9, pp. 253-306.

Ramsey, Arthur Stanley (2009). *Electricity and Magnetism: An Introduction to the Mathematical Theory*. Cambridge: Cambridge University Press.

Reich, S. (2011). Edwin Hubble in translation trouble. *Nature* (Online).

Rescher, N. (1967). *The Philosophy of Leibniz*. Englewood Cliffs, NJ: Prentice-Hall.

Restivo, Sal (2005). *Science, Technology, and Society: An Encyclopedia*. New York: Oxford University Press.

Rheinberger, H. (2005). Gaston Bachelard and the Notion of 'Phenomenotechnique'. *Perspectives on Science* 13(3): pp. 313-328.

——— (2010). *On Historicizing Epistemology: An Essay* (D. Fernbach, Trans.). Stanford, CA: Stanford University Press.

Robinson, Ronald, Gallagher, John, and Alice Denny (1961). *Africa and the Victorians: The Climax of Imperialism in the Dark Continent*. New York: St. Martin's Press.

Rochberg, Francesca (1999). Empiricism in Babylonian Omen Texts and the Classification of Mesopotamian Divination as Science. *The Journal of the American Oriental Society* 119(4): pp. 559-569.

Rocher, Ludo (1986). The Puranas. In: *A History of Indian Literature*, edited by Jan Gonda, Vol. II, Fasc. 3. Wiesbaden: Otto Harrassowitz.

Rose, Steven (2005). *The Future of the Brain: The Promise and Perils of Tomorrow's Neuroscience*. New York: Oxford University Press.

Rosenberg, A. (2000). *Philosophy of Science: A Contemporary Introduction*. London: Routledge.

Rosenblatt, Frank (1962). *Principles of Neurodynamics; Perceptrons and the Theory of Brain Mechanisms*. Washington: Spartan Books.

Rosenfeld, B.A. (1988). *A History of Non-Euclidean Geometry. Evolution of the Concept of a Geometric Space*. Berlin: Springer.

Roth, Cecil (1977). *Doña Gracia of the House of Nasi*. Philadelphia: Jewish Publication Society.

Rothenberg, Marc (Ed.) (2001). *The History of Science in the United States: An Encyclopedia*. New York: Garland.

Rüegg, Walter (Ed.), (1993). *Geschichte der Universität in Europa*, 3 Vols. München: C.H. Beck.

Russell, Stuart J., and Norvig, Peter (Eds.). (2010). *Artificial Intelligence: A Modern Approach* (3rd ed). Upper Saddle River, NJ: Prentice Hall.

Saïd, S., Trede, M., de Boulluec, A. (2001). *History of Greek Literature*. Athens: Papazisis.

Samuelson, Robert. J. (2006). The Next Capitalism; American Business Is in the Midst of Its Greatest Transformation since the Industrialization and Massive Growth at the Turn of the 20th Century. *Newsweek*, October 30: p. 45.

Sarton, G. (1952). *A Guide to the History of Science: A First Guide for the Study of the History of Science, with Introductory Essays on Science and Tradition.* Waltham, MA: Chronica Botanica.

———— (1957). *The Study of the History of Mathematics, and the Study of the History of Science.* New York: Dover Publications.

———— (1959). *Ancient Science and Modern Civilization. Euclid and His Time. Ptolemy and His Time. The End of Greek Science and Culture.* New York: Harper.

———— (1960). *The Life of Science: Essays in the History of Civilization.* Bloomington: Indiana University Press.

Saudi Arabia: Government makes clear that it now needs foreign investment for research, development and production (1999). *Oil*, 139: p. 5.

Saunier, Pierre-Yves (2009). Transnational. In: Akira Iriye et Pierre-Yves Saunier, *The Palgrave Dictionary of Transnational History*, Palgrave Macmillan: pp. 1047-1055. <halshs-00368360>

Scamehorn, Lee (2002). *High Altitude Energy: A History of Fossil Fuels in Colorado.* Boulder, CO: University Press of Colorado.

Scarani, Valerio (2006). *Quantum Physics: A First Encounter: Interference, Entanglement, and Reality.* Translated by Rachel Thew. New York: Oxford University Press.

Schafer, Edward H. (1963). *The Golden Peaches of Samarkand: A Study of T'ang Exotics.* Berkeley and Los Angeles: University of California Press.

Schaffer, Simon (1994). Babbage's Intelligence: Calculating Engines and the Factory System. *Critical Enquiry* 21: pp. 203-227.

Schehr, Robert C. (1997). *Dynamic Utopia: Establishing Intentional Communities as a New Social Movement.* Westport, CT: Bergin and Garvey.

Schmidgen, H. (2015). *Bruno Latour in Pieces: An Intellectual Biography* (G. Custance, Trans.). New York: Fordham University Press.

Schmidt W. (transl.) (1899-1914). *Heronis Alexandrini opera quae supersunt omnia.* Vol. 1: *Die Druckwerke Herons von Alexandria.* Leipzig: Teubner.

Schott, Heinz (2002). Paracelsus and Van Helmont on Imagination: Magnetism and Medicine before Mesmer. In: *Paracelsian Moments: Science, Medicine and Astrology in Early Modern Europe* (Edited by Gerhild Scholz Williams and Charles D. Gunnoe Jr.). Kirksville, MO: Truman State University Press, 2002: pp. 135-147.

Schroeder, Ralph (1996). *Possible Worlds: The Social Dynamic of Virtual Reality Technology.* Boulder, CO: Westview Press.

Shuren, Bo (1989). Astrometrie und astrometrische Instrumente. In: *Wissenschaft und Technik im alten China* (Edited and translated by Käthe Zhao and Hsi-lin Zhao). Berlin: Birkhäuser: pp. 20-36.

Schuld, Maria, Sinayskiy, Ilya, and Francesco Petruccione (2015). An introduction to quantum machine learning. *Contemporary Physics* 56(2): pp. 172-185.

Scrimshaw, Peter (Ed.). (1993). *Language, Classrooms and Computers.* London: Routledge.

Segal, Howard P. (1994). *Future Imperfect: The Mixed Blessings of Technology in America.* Amherst, MA: University of Massachusetts Press.

Serres, M. (1968). *Le Systeme de Leibniz et ses Modeles Mathématiques, Tome Premier, Etoiles.* Paris: Presses Universitaires de France.

Sezgin, Fuat (2003). *Wissenschaft und Technik im Islam.* Frankfurt am Main: Institut für Geschichte der Arabisch-Islamischen Wissenschaften an der Johann Wolfgang Goethe-Universität.

Sexton, Donal J., Jr. (comp.). (1996). *Signals Intelligence in World War II: A Research Guide.* Westport, CT: Greenwood Press.

Shanker, S. G. (Ed.). (1996). *Philosophy of Science, Logic, and Mathematics in the Twentieth Century.* New York: Routledge.

Sherby, Oleg. D. and Jeffrey Walsworth (2001). Ancient Blacksmiths, the Iron Age, Damascus Steels, and Modern Metallurgy. *Journal of Materials Processing Technology* 117: pp. 347-353.

Silverberg, R. (1997). *The longest voyage: Circumnavigators in the Age of Discovery.* Athens, OH: Ohio University Press.

Simmons, Matthew R. (2005). *Twilight in the Desert. The Coming Saudi Oil Shock and the World Economy.* Hoboken, NJ: Wiley.

Singh, G. (2009). *Applied Chemistry.* New Delhi: Discovery Publishing House.

Skelton, Raleigh A. (1958). *Explorers' maps: Chapters in the cartographic record of geographical discovery.* New York: Frederick A. Praeger.

Sklar L. (1974). *Space, Time and Spacetime.* Berkeley: University of California Press.

Smith, Cyril S. (1943). Introduction. In: Vanoccio Biringuccio, *Pirotechnia.* New York: The American Institute of Mining and Metallurgical Engineers.

Smith, Elisabeth B. and Michael Wolfe (ed.). (1997). *Technology and Resource Use in Medieval Europe.* Aldershot: Ashgate.

Smithsonian Institution (1880). *A Memorial of Joseph Henry.* Washington, DC: Government Printing Office.

Smuts, Aaron (2009). What is Interactivity? *The Journal of Aesthetic Education,* 43(4).

Sobel, Dava (1995). *Longitude: The True Story of a Lone Genius who Solved the Greatest Scientific Problem of his Time.* London: Penguin.

Spandagos, Evangelos (2000). *Theodosius' Sphaerics.* Athens: Aithra.

Stanney, Kay M. (Ed.). (2002). *Handbook of Virtual Environments: Design, Implementation, and Applications.* Mahwah, NJ: Lawrence Erlbaum Associates.

Steels, Luc, and Rodney Brooks (1995). *The Artificial Life Route to Artificial Intelligence: Building Embodied, Situated Agents.* Hillsdale, NJ: Lawrence Erlbaum Associates.

Sterling, Keir B., Richard P. Harmond, George A. Cevasco, and Lorne F. Hammond, (Eds.). (1997). *Biographical Dictionary of American and Canadian Naturalists and Environmentalists.* Westport, CT: Greenwood Press.

Sternberg, Robert J., and Talia Ben-Zeev (Eds.) (1996). *The Nature of Mathematical Thinking.* Mahwah, NJ: Lawrence Erlbaum Associates.

Stinson, Douglas R. (2006). *Cryptography: Theory and Practice.* Boca Raton, FL: Chapman and Hall/CRC.

Tabbi, Joseph, and Rone Shavers (Eds.). (2007). *Paper Empire: William Gaddis and the World System.* Tuscaloosa, AL: University of Alabama Press.

Takeyh, Ray (2000). *The Origins of the Eisenhower Doctrine: The US, Britain and Nasser's Egypt, 1953–57.* London: Palgrave Macmillan.

Taylor, Andrew (2004). *The World of Gerard Mercator: The Mapmaker who Revolutionised Geography.* London: William Collins.

Taylor, Frederick Winslow (1917). *The Principles of Scientific Management.* New York: Harper and Brothers.

Teaf, Howard M. Jr., and Peter G. Franck (1955). *Hands across Frontiers: Case Studies in Technical Cooperation.* Ithaca, NY: Cornell University Press.

Teays, Wanda (2012). *Seeing the Light: Exploring Ethics through Movies.* Malden, MA: Wiley-Blackwell.

Tebel, René (2012). *Das Schiff im Kartenbild des Mittelalters und der Frühen Neuzeit. Kartographische Zeugnisse aus sieben Jahrhunderten als maritimhistorische Bildquellen.* Bremerhaven: Deutsches Schiffahrts Museum, Oceanum Verlag.

Thang, Leng Leng, and Wei-Hsin Yu (Eds.). (2004). *Old Challenges, New Strategies: Women, Work, and Family in Contemporary Asia.* Boston: Brill.

Thompson, J. W., F. Schevill, G. Sarton, and G. Rowley (1929). *The Civilization of the Renaissance.* Chicago: University of Chicago Press.

Thucydides (1970). *Historiae.* Oxford Classical Texts. Athens: Kardamitsa.

Topper, David R. (2015). *Einstein for Anyone: A Quick Read.* Delaware: Vernon Press.

Trinkle, Dennis A. (Ed.). (1998). *Writing, Teaching, and Researching History in the Electronic Age: Historians and Computers.* Armonk, NY: M. E. Sharpe.

Tropp, Edward A., Victor Ya. Frenkel and Artur D. Chernin (1993). *Alexander Friedman: the man who made the universe expand* (Translated by Alexander Dron and Michael Burov). Cambridge: Cambridge University Press.

Tsimpourakis, Dimitris (2004). *The Geometry in Ancient Greece.* Athens: Atrapos.

Tweney, C.F. and L.E.C. Hughes (1958). *Chamber's Technical Dictionary.* New York: Macmillan.

Tyndall, John (1961). *Faraday as a Discoverer.* New York: D. Appleton.

U.S. Congress Office of Technology Assessment (1984). *Computerized Manufacturing Automation: Employment, Education, and the Workplace.* Washington, DC: U.S. G.P.O.

Unguru, Sabetai (1975). On the need to rewrite the history of Greek mathematics. *Archives for the History of Exact Sciences* 15: pp. 67–114.

Uzor, O.O. (2004). Small and Medium Scale Enterprises Cluster Development in South-Eastern Region of Nigeria. *Berichte aus dem Weltwirtschaftlichen Colloquium der Universität Bremen*, Nr. 86. Institut für Weltwirtschaft und Internationales Management. Universität Bremen.

Valenstein, Elliot S. (2005). *The War of the Soups and the Sparks: The Discovery of Neurotransmitters and the Dispute over How Nerves Communicate.* New York: Columbia University Press.

Van den Boogaerde, Pierre (2010). *Shipwrecks of Madagascar.* Durham: Strategic Book Group.

Van der Waerden, B.L. (2003). *Science Awakening I. Egyptian, Babylonian and Greek Mathematics* (2nd Ed.). Irakleio: Crete University Press.

Van Heijenoort, Jean (1967). *From Frege to Gödel: A Source Book in Mathematical Logic, 1879-1931.* Cambridge, MA: Harvard University Press.

Vassiliou M.S. (2009). *Historical Dictionary of the Petroleum Industry.* Lanham MA: Scarecrow Press.

Véron, Jacques (2008). Alfred J. Lotka and the Mathematics of Population. *Electronic Journal for History of Probability and Statistics* 4(1), http://www.jehps.net/juin2008.html

Vestine, E.H., Lucile Laporte, Isabelle Lange, and W.E. Scott (1947). *The Geomagnetic Field: Its Description and Analysis.* Washington, DC: Carnegie Institution of Washington.

Virtanen, R. (1960). *Claude Bernard and His Place in the History of Ideas.* Lincoln, NE: University of Nebraska Press.

Vitrac, Bernard (2008). *Faut-il réhabiliter Héron d'Alexandrie?* France.

Vitruvius (1914). *The Ten Books on Architecture.* London: Oxford University Press.

Vlastos, Gregory (2005). *Plato's Universe*. Las Vegas, NV: Parmenides.

Vogl, Benedikt (2014). *Die Amerikapolitik Karls V.* Diplomarbeit, University of Vienna, Historisch-Kulturwissenschaftliche Fakultät.

Volti, Rudi (2014). *Society and Technological Change* (7th Ed.). New York: Worth Publishers.

Von Humboldt, Alexander (1858). *Cosmos: A Sketch of Physical Description of the Universe*, vol. 1. Translated by E.C. Otté. New York: Harper and Brothers.

Von Reitzenstein, Alexander (1959). Die Ordnung der Nürnberger Plattner. *Waffen- und Kostümkunde*, New Series, I: 54-85. München.

Voth, Hans-Joachim (2000). *Time and Work in England 1750-1830*. New York: Oxford University Press.

Wagman, Morton (Ed.). (1991). *Cognitive Science and Concepts of Mind: Toward a General Theory of Human and Artificial Intelligence*. New York: Praeger Publishers.

Wallerstein, Immanuel (1974). *The Modern World System: Capitalist Agriculture and the Origins of the European World-Economy in the Sixteenth Century*. New York: Academic Press.

Weiss, R. (2001). *Virtue in the Cave. Moral Inquiry on Plato's Meno*. Oxford University Press.

West, Delno C., and August Kling (trans.) (1991). *The Libro de las Profecias of Christopher Columbus*. Gainesville, FL: University of Florida Press.

Westfall, Richard S. (2004). *The Construction of Modern Science. Mechanisms and Mechanics* (4th Ed.), (Trans. Krino Zisi). Irakleio: Crete University Press

Weston, R. F. and Ruth, M. (1997). A dynamic, hierarchical approach to understanding and managing natural economic systems. *Ecological Economics* 21: pp. 1 - 17.

Wheeler, John Archibald (1998). *Geons, Black Holes, and Quantum Foam: A Life in Physics*. New York: W. W. Norton.

Whittaker, John C. (2004). *American Flintknappers: Stone Age Art in the Age of Computers*. Austin, TX: University of Texas Press.

Wieland, H. R. (2013). *Computergeschichte(n) – nicht nur für Geeks. Von Antikythera zur Cloud*. Bonn: Galileo Press.

Wilcken, U. (1976). *Ancient Greek History*. Athens: Papazisis.

Wilhelm, Richard (1979). *Lectures on the 'I Ching': Constancy and Change*. Princeton NJ: Princeton University Press.

Williams, Alan (2003). *The Knight and the Blast Furnace: A History of the Metallurgy of Armour in the Middle Ages and the Early Modern Period*. Leiden, Boston: Brill.

Williams, Garnett P. (1997). *Chaos Theory Tamed*. London: Taylor and Francis.

Williams, Gerhild Scholz, and Charles D. Gunnoe Jr. (Eds.). (2002). *Paracelsian Moments: Science, Medicine and Astrology in Early Modern Europe.* Kirksville, MO: Truman State University Press.

Williamson, Harold F., and Arnold R. Daum (1959). *The American Petroleum Industry: The Age of Illumination, 1859-1899.* Evanston, IL: Northwestern University Press.

Wilson, N. L. (1973). Individual Identity, Space, and Time, in the Leibniz Clarke Correspondence. In: Leclerc I. (Ed.). *The Philosophy of Leibniz and the Modern World,* Vanderbilt University Press, Nashville: pp. 189-206.

Winterbourne, A. T. (1982). On the Metaphysics of Leibnizian Space and Time. *Studies in the History and Philosophy of Science* 13(3): pp. 201-14.

Yi, Dongshin (2010). *A Genealogy of Cyborgothic: Aesthetics and Ethics in the Age of Posthumanism.* Farnham, Surrey, England: Ashgate.

Yongming, Zhou (2006). *Historicizing Online Politics: Telegraphy, the Internet, and Political Participation in China.* Stanford, CA: Stanford University Press.

Youngman, Paul A. (2010). The Dada Cyborg: Visions of the New Human in Weimar Berlin. *German Quarterly.*

Yourgrau, Palle (2005). *A World without Time: The Forgotten Legacy of Gödel and Einstein.* Cambridge MA: Basic Books.

Zahar, Elie (2001). *Poincaré's Philosophy: From Conventionalism to Phenomenology.* Chicago: Open Court.

Index

www.ingramcontent.com/pod-product-compliance
Lightning Source LLC
Chambersburg PA
CBHW070716220326
41598CB00024BA/3187